Leave No
THRESHING STONE
Unturned

Leave No Threshing Stone Unturned / by Glen Ediger, 1953

Library of Congress: 2012914593

ISBN: 978-0-615-68201-3

Includes bibliographic references

Printed in the United States of America

Printing Inc. Wichita, Kansas

Design: Glen Ediger

Layout: Glen Ediger

Cover Design: Glen Ediger

Written by: Glen Ediger

Edited by: Dr. Rachel Pannabecker

Published by: Glen Ediger - 14 Regal Crescent - North Newton, Kansas 67117

Published: September, 2012

EDIGER DESIGN

Visit on the internet @ www.threshingstone.com

Acknowledgments

This book is the result of basic curiosity...curiosity led me to ask questions...questions led to conversations...conversations led to research...research led to quests...and quests led to lists.

Eventually I had to decide what to do with all this information. With the encouragement of historians, authors, friends, and family I decided to collect this information into the form of a book.

It is a very esoteric group of individuals who have even heard of a threshing stone, yet despite its obscurity, I think it is a story that is worthy of being told.

With the paths that the threshing stone crosses, I think many will find the stories both interesting and educational. I hope people who love history, farming, wheat, railroads, Kansas, the prairies, Bethel College, Mennonites, Russian-German immigrants, and bread, find some satisfaction in perusing these pages.

I could not have done this book without the help and tolerance of my wife Karen, whose companionship and encouragement through the years have made me who I am today. We have enjoyed many day trips, meeting wonderful people throughout this process. I also appreciate her assistance in the conceptual work and for first draft editing.

History was not always one of my top interests, but as I matured through the years, the stories and artifacts of the past have piqued my interests. My involvements with Kauffman Museum started when I was attending Bethel College and worked one summer helping to move the Voth-Unruh-Fast House to North Newton. I had the privilege of working with Cornelius Krahn and Ernie Unruh, sweating hard and learning much through that experience. My involvement with Kauffman Museum continues as a volunteer and being the President of the Board for the last 3 years.

I want to say thanks to all the historians that have documented history and informed me through their writing, personal discussions, and their collection of artifacts. Thanks in particular to those historians that I have personally known; Cornelius Krahn, John Thiesen, Ray Wiebe, Rachel Pannabecker, Peggy Goertzen, D. C. Wedel, Harley Stucky, Steve Friesen, Jim Juhnke, Keith Sprunger, Kelly Harms, Jerry Toews, Leann Toews, Mil Penner, Marci Penner, Virgil Litke, Karen Penner, Jay Price, Robert Collins, John Schmidt and David Haury.

Thanks to Bob Regier for his mentoring throughout the years; his design influence helped me throughout my career. Thanks to Harold Thieszen and Ray Regier for translation assistance, thanks to Laurie Schrag, Rebecca Epp, John Thiesen and Karen Ediger for help in editing. Thanks to Paul Johnston and Will Gilliland for advice on geological research. Particular thanks to John Thiesen and James Lynch for help in finding records at the Mennonite Library and Archives and a special, big thanks, to Dr. Rachel Pannabecker for editing.

Thanks to my family, my parents Albert M. and Lydia Ediger, along with my brothers LaVon and Byron Ediger, for sharing their lives, knowledge, and values as we grew up together on our family farm near Buhler, Kansas. Thanks also to my son Brandan Ediger for all the times we have spent together; he is a very special person in my life.

My final acknowledgment goes to those stone cutters, farmers, and agronomists that evolved the processes of their crafts through the years and created this obscure and curious tool.

Threshing Stone on display at Alexanderwohl Mennonite Church - Goessel, Kansas.

Finding The Stones

On most pages throughout this book, there is a printed texture of the threshing stone surface, as in this area.

In this area is a running account of the quest to find the threshing stones and the stories that go with them.

This is a parallel story that is not directly tied to the various articles printed on the white portion of the pages.

See Article 17 for the methods of finding the stones.

I hope you enjoy this unique format and feel free to bounce back and forth between the parallel texts as you desire.

Names of owners used by permission.

Contents

Foreword

Who am I and why would I write a book about threshing stones in Kansas? I don't think the story of the threshing stones is just a boring story about rocks, but rather that it is a fascinating story, a story that crosses the paths from many directions at a specific time in history.

 Little did I know when I decided to take on this project that it would be so complex, so diverse, or so interesting. But I find this story to be rich with the intertwining of world politics, economics, ethnicity, religion, agronomy, archeology, world food supplies, the railroad, the Wild West, displacement, and opportunity.

But for the slimmest chance in history, there would be no threshing stones in North America. As I researched this rare farming artifact I was amazed at how unique and significant it is.

I suspect this farming relic would be lost to history if it had not been given great notoriety by being adopted as the mascot for Bethel College in North Newton, Kansas. If not for the stone that was made for H. Richert of the Alexanderwohl community, which was then passed to his son-in-law C. H. Wedel, the first president of Bethel College, then placed in an upright position in front of his house in 1903, and that stone eventually ending up on the Bethel College campus, it may have never become the symbol of Bethel College on November 16, 1934.

If not for these events, I believe these outdated farming tools would have fallen into obscurity and I would not have had a personal connection or interest in writing this book. But many stories and facts have been discovered as I have tried to find the layered history of

this simple limestone farming tool. As an agricultural artifact, the threshing stone has a place in the long history of grain production, visible throughout the world. The threshing stone represents a unique blend of attributes; this story is about a stone, a tool, a symbol, and a heritage.

-I am intrigued with it just as a "Stone" that was carved from native bedrock and shaped into a piece of functional artistry by the hands of skilled stone cutters, creating the tool that is today only a visual reminder of our past heritage.

-I am intrigued with it as a "Tool" that has its place in agricultural history that started with manual labor using only hooves and flails for threshing, then to the use of the threshing stone to reduce labor and then becoming obsolete with the invention of the mechanical threshing machine.

-I am intrigued with it as it traces the history of a "Heritage," in particular the story of the Mennonite farmers as they migrated with their families and farming skills from country to country for religious freedom, ultimately bringing the hard red winter wheat to the plains of Kansas.

-I am also intrigued with the threshing stone as a "Symbol," adopted by Bethel College, where it became the symbol of strength and endurance, rooted in the hard working farm families of the plains.

I found this story worthy of in-depth research that has culminated in this book, but why did I do it? I have many reasons for taking on this project, learning as much as possible about the history and use of the threshing stone. I have as many reasons as there are facets on a stone.

Our Threshing Stone.

1. *I'm a Kansas farm kid*
I grew up on a wheat farm in western Harvey County on Dutch Avenue, so named for all the "Dutch" ancestry Mennonites that settled along this road when they migrated to Kansas. My roots are in farming; our wheat farm was the center of my universe for much of my life. Everything was somewhere from there. The farm is etched deep in my psyche, and even though I have not farmed for decades, in my heart I am still a farmer.

2. *I'm an industrial designer*
I have designed products for the mass market my entire career. As Director of Design for Vornado Air LLC I have been involved in all of Vornado's products for over 2 decades. Because of my career in product development, my interest was piqued by the design of the threshing stone, how it worked, how it was made, and asking, "Why did this product exist?"

3. *I'm a patron of history*
I am not a professional historian but I have been involved in historical ventures my whole life. I am currently President of the Board of Directors of Kauffman Museum, I am a past board member of the museum of the Antique Fan Collectors Association, and I get a lot of satisfaction from the antiques that my wife and I have inherited and collected over the years.

4. *I'm a Mennonite*
My ancestors immigrated to Kansas from the Ukraine in 1874. They came to Kansas at the invitation of the State of Kansas and the Santa Fe Railroad, in order to escape their loss of religious freedoms and to develop the prairie into a rich and productive agricultural economy that could support their family. These groups of families brought with them the Turkey Red winter wheat, along with their farming skills, hard work ethic, and traditions.

5. *I'm a Bethel College graduate*
I had a wonderful time as a student at Bethel where the threshing stone is the symbol of the college. I suspect that virtually every student that has attended Bethel is aware of the threshing stone.

6. *We own a threshing stone*
My wife inherited the stone from her paternal grandmother's family from the Goessel community. This stone got me asking questions and started my interest in this project.

7. *No one had done it yet*
I was surprised that no one had done any in-depth research on threshing stones, so why shouldn't I? I live in the middle of the Kansas prairie, I have deep connections to the people of this area, I have numerous friends that are writers and historians, and I live just blocks away from the Bethel campus, the Mennonite Library and Archives, and Kauffman Museum.

When we got our threshing stone, I was very proud to be an owner of one of these historic relics, but I didn't know much about them. I asked individuals who are historians and antique collectors as to what they knew, who made the stones, and how they actually were used. But few had any answers. I had always assumed there must be thousands spread out across North America.

Author at Kansas State Historical Society Archives Research Room.

Several individuals had speculated that there may only be 100 stones. I was amazed that the number could be so low and was doubtful if that could be true. But the more I learned, the more certain I thought that they might be right.

Even though the majority of people in the world have no idea what a threshing stone is, it is still an important symbol and artifact, and known by many Mennonites and others around the world.

For all these reasons I decided to take on the task of learning more. The first person I called was Brian Stucky, an art teacher from Goessel, Kansas, a historian, an author, a college classmate, and a friend. He and I had talked about threshing stones over the years and he had even started to compile a list of stones that he knew of years earlier. I asked him if he was going to pursue additional research on threshing stones, and told him that I had started to compile a bunch of information and that I would be glad to assist him in his research, or if not, I would like to take it on as a project and I would appreciate his help. He said he was currently involved in other projects and encouraged me to go for it. So I did.

My questions were simple at first; how many existed, how wide spread was their use, where were they made, and how did they actually work. Little did I know how complex the answers would become.

Very quickly I also realized I had to try to see how many I could find. Again, little did I know how complex or fun this would be. I decided first to just see how many I could find.

I hope you enjoy the stories about my quest to find the stones and the stories that I was told about those I found. To say

the least, this project has been incredibly fun. I never knew I could have so many hours of enjoyment in archives, libraries, used book stores, and museums. Finding the stones was like a big treasure hunt. I didn't know how to start, where to look, or what I might find, but figuring that out was part of the fun.

Most of all, I appreciated the many wonderful conversations I've had with those individuals who still have threshing stones in their families and hearing the stories that go with them.

My goal in compiling this book is first to present the complete history of the threshing stone that will stand up to academic scrutiny, and also tell the story in an interesting and enjoyable narrative. The format I have chosen is somewhat like a magazine - pick up and read any article. It is not necessary to read it in sequence, but this creates some redundancy. I have gone to great depths to cover each facet of the story in detail (possibly way too much detail at times) to assure that the information is historically correct and complete as I know it today.

History is written with the bias of the author. I expected the threshing stone story to be spread broadly across North America and over many ethnic groups, but I found that this story is primarily about central Kansas Mennonites. If I have overlooked any significant resources or seem overly focused on my home turf, it is because the research has led me back to my roots.

I expect my quest will not be over at the publishing of this book but additional information and stone finds will continue, see updates at www.threshingstone.com. I hope that you find this book not only informative, but also fun to read and to share with others.   ENJOY!

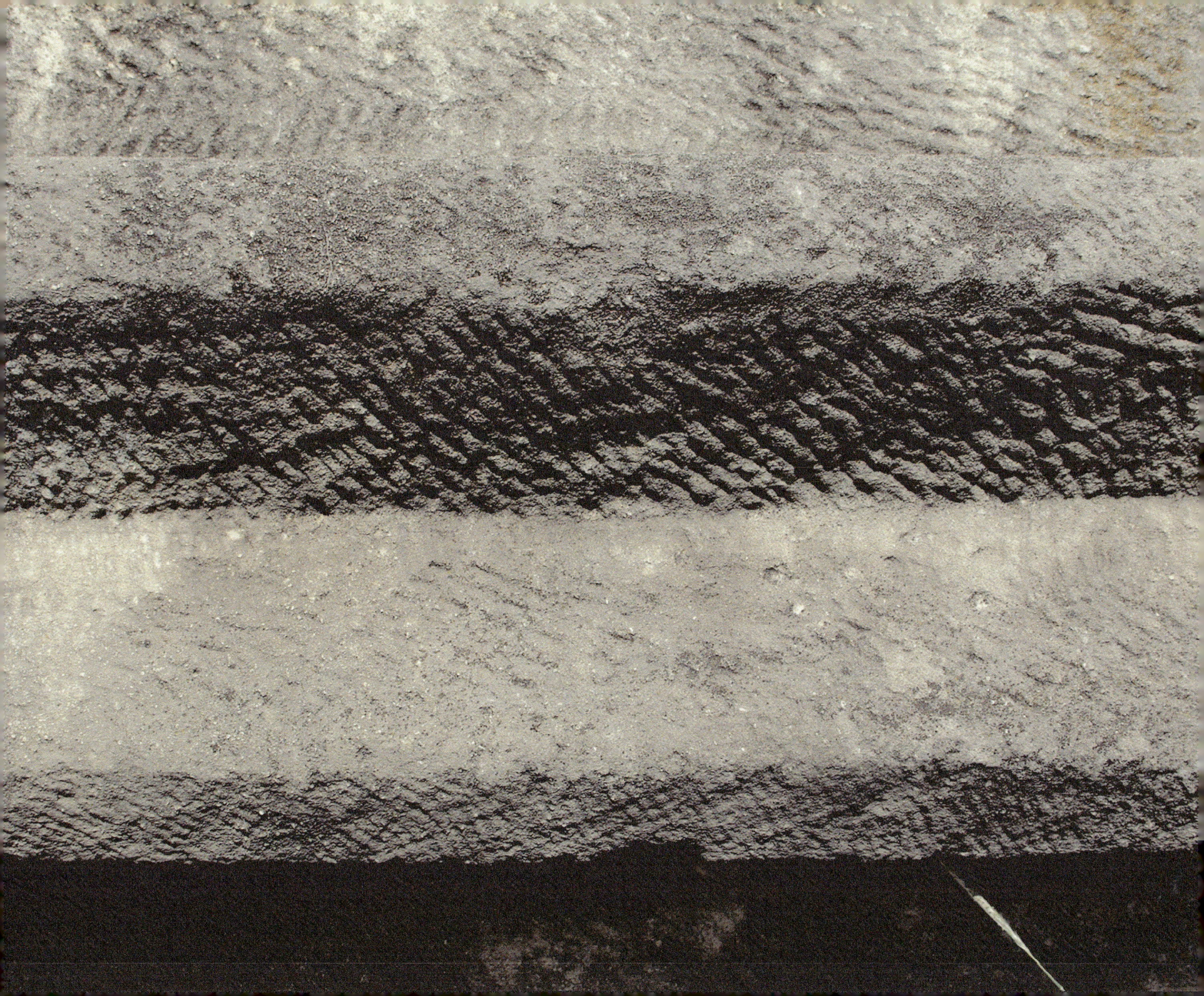

Threshing Stone Basics

The tilling of the ground, the planting of the seeds, the methods of reaping, and the process of separating the grain from the straw have evolved throughout the years. In this article I will focus on the basic story of the threshing stone, the process of threshing the grain, and how the threshing stone came to be used among Mennonites in Kansas.

The story of the simple threshing stone is actually a very complex one. The treasured artifact of farming history would be all but forgotten had it not been symbolized and preserved by those who are lucky enough to still have one in their possession. So what is a threshing stone, how was it used, and why do we care?

If not for the timely crossing of many world events at a specific time in history, there would be no threshing stones in North America. Many events that included international politics, the opening of the Wild West, economic opportunities, the railroad, the search for religious freedoms, and changes in farming technologies, all crossed paths in the 1870s to create the environment where there existed the brief use of threshing stones in North America.

Some have asked if it is a story of failure or a story of success. As a tool the threshing stone had its place in time, an improvement over the much more laborious methods of threshing grain, but a tool that eventually came to the end of its time for practical use. However I consider it a symbol of endurance and ingenuity, a tool used to thresh the wheat from the chaff, and as a symbol of human capacity to explore and distinguish what is good from what is bad. It is a symbol that reminds us of a rich heritage, steeped in values of farming families, pioneers, religious seekers, and entrepreneurs. It is a wonderful symbol.

It starts with the need for food. The growing of cereal grains has evolved over the last 10,000-12,000 years that humans have been using grain as a

Opposite - Threshing in Russia - circa 1900.
Mennonite Library and Archives. MLA 2004-0153

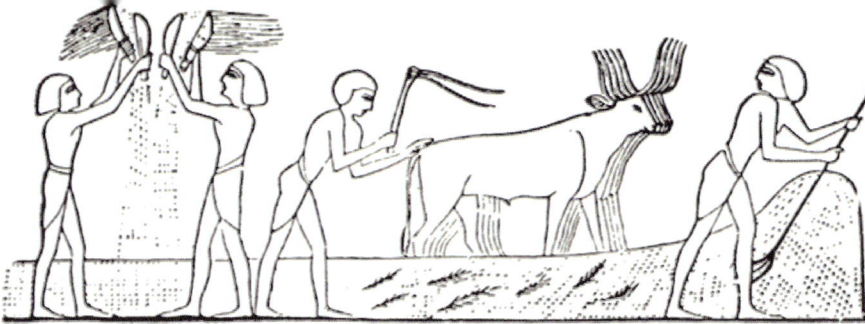

Egyptian Threshing and Winnowing - Harold B Hunting.
From: http://www.gutenberg.org/files/18187/18187-h/18187-h.htm#imagep44b

food source. There is little doubt that the ability of humans to cultivate, grow, process and turn grain into food was one of the most profound turning points in human history. With the ability to cultivate food, rather than just follow animals to hunt

Unleavened Bread - Puri
Ashok Modhvadia - 2009.
Wikimedia Commons

for food, groups of humans could remain in one place, no longer needing to follow the food source. This development in human skills allowed for the creation of communities, which could finally function in new and different ways. People no longer needed to be mobile,

allowing individuals to specialize in trades such as farming and allowed others to provide skilled services to the community. Now civilization could depend on the farmers to provide food for the community throughout the year. This inspired trade as well as increasing the need for communication and the capability to plan and save for the future. Farming changed civilization. *[See Article 2 for more in-depth details about cereal grains.]*

Farming changed very little from early times until about 1700. Then an agricultural revolution took place which led to a large increase in the production of crops. In the 1850s, the industrial revolution spilled over to the farm with new mechanized methods which increased production rates. Large changes in the use of new tools, combined with crop rotation, use of manure, and better soil preparation, led to a steady increase of crop yield in Europe.[1]

In the mid 16th century, many Mennonites and other religious groups pursued religious freedoms and escaped persecution from various European locations. Mennonites looked for new opportunities, traveling from Switzerland, Northern Germany, and the Netherlands to places like Prussia (now Poland), living there for about two hundred years. They did wonders with farming practices but

Threshing Stone #1

This is the stone that inspired me to research threshing stones. This stone now sits in our yard in North Newton, Kansas. Time and weather have taken a toll on this old stone and it has unfortunately broken into 3 pieces.

The stone comes from Karen's father, Ernest G. Unruh. Ernest's mother was Elizabeth Goertz and this stone was from her family. Their home farm was located 1 mile west of K-15 and north of Goessel about 4.5 miles on the east side. Karen's cousin, Darrel Unruh gave us an old photograph of Karen's grandmother standing by the stone in front of the house on the original farm, taken in about 1939. We do not know any of the stories prior to that date, but assume that it is original to this farm from about the time of their immigration in 1874.

Karen recalls that in the 1970s the stone was moved from that farm to the farmstead of Ernest and Ruth (Schmidt) Unruh on the south side of Goessel. This was the home place of Karen's mother, often referred to as the farm by the old Gordon Schoolhouse.

This stone sat upright in their yard east of the summer kitchen for about 20 years. Unfortunately at some time part of the stone broke and was held together with a steel ring around the top. When they moved to a duplex apartment

Oldest known photograph of a Mennonite threshing stone in use, Circa 1890s, Russia. MLA - 2004-0080

once again, fearing the loss of religious freedoms and the need for new economic opportunities, looked to find a better country. *[See Article 7 for more in-depth information about Mennonite migrations.]*

At the invitation of Catherine the Great, beginning in 1789, many families and entire communities relocated to the Russian Ukraine, to develop the fertile steppes into rich and productive farm ground. Wheat soon became one of the main crops grown in this area. *[See Article 12 for more in-depth details on wheat.]*

Evidence of the use of threshing stones in various forms dates from the Roman era, but eventually it became the preferred method

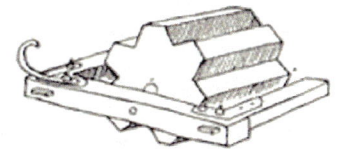

Ukrainian Threshing Stone - ethnography.org.ua, Article by V. Gorlenko - "Types of Economic Activity"

of threshing grain by Mennonite farmers in the Ukraine for about 60 years. Although at this time threshing stones were not exclusive to Mennonites, the Mennonite farmers were the primary users of this process.

Threshing with a stone was the method used to knock the grain from the head because it was less labor-intensive than using the traditional flail. I have found evidence of hundreds of threshing-stone remains throughout the Ukraine, either abandoned in a ditch to suffer the consequences of time, or saved to be found in museums and monuments

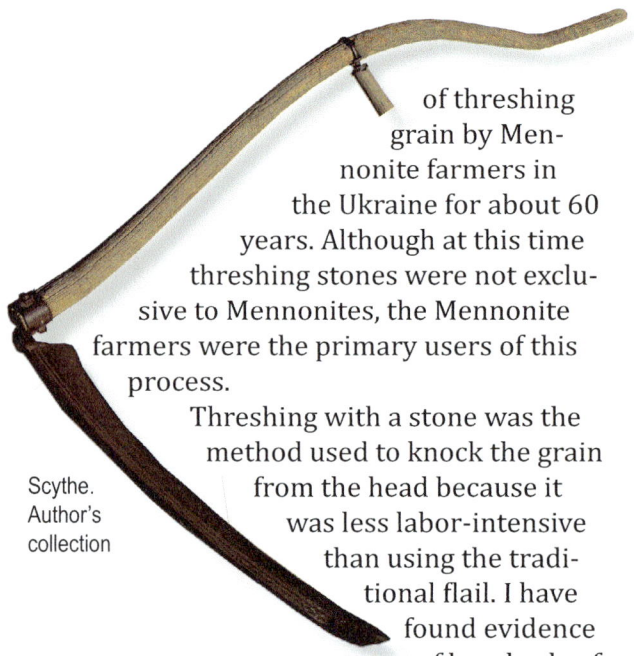

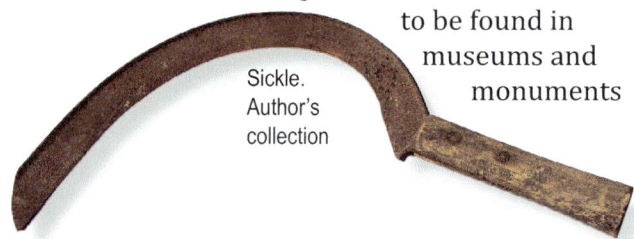

Scythe. Author's collection

Sickle. Author's collection

as a testament to their contribution to a culture. Early Ukrainian threshing stones were made of logs, with longitudinal ribs; later they were carved from native sandstone and granite, but late 19th century designs were made of concrete.[2]

Shocks of wheat in field by Goessel, Kansas - 2012.

Threshing Stone located in the Molotschna colony in the Village of Grosswelde. Lowell Ratzlaff 1997

[See Article 16 for more in-depth details about finding the stones.]

In the Ukraine in the 1800s the entire process of harvesting, threshing, separating and storing the grain, whether it was rye, barley, or wheat was very labor-intensive. The grain needed to be harvested at the proper time to insure that the seed was full and developed in the head. Early reaping was all done by hand. The straw

Russian Cradle Scythe. Kauffman Museum

at Bethesda Home of Goessel, we moved it to their back yard close to the entrance of the Goessel Mennonite Heritage Museum where it sat for several years.

Eventually we moved the stone to its present location at our home in North Newton, where we treasure its legacy, appreciate having it and are excited that it provoked my curiosity. We decided to remove the steel ring and let it be shown in pieces. Some have said how sad it is that it broke, but I still appreciate it as a historic visual symbol and think that being in pieces adds to its visual intrigue.

Threshing Stones 2-4

The threshing stones on the Bethel Campus are a curious bunch. These stones have graced the campus, some for over a hundred years, and are at least part of the reason we even know what a threshing stone is today. Though the threshing stone has no exclusive claim by Bethel and has been used in many communities, had they not been brought to the campus and then to eventually become the symbol of the college, I sus-

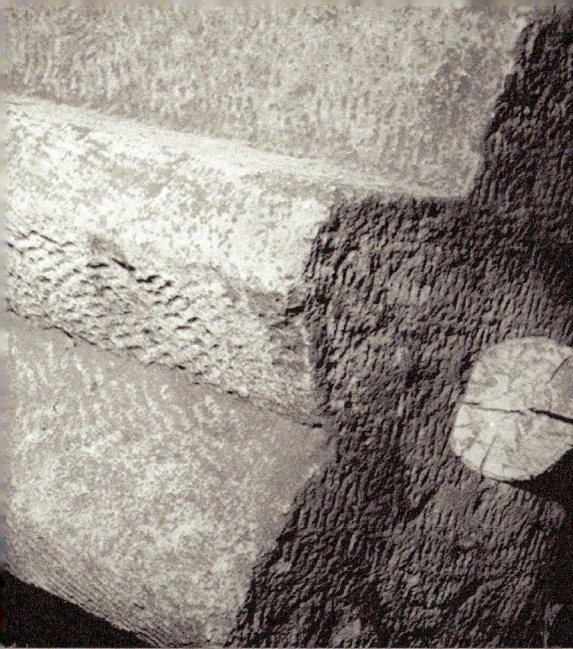
1959 close-up of original campus stones. MLA

pect that only antique collectors would know of these rare artifacts today.

A bit of Bethel history as noted by John Thiesen in his lecture as to how the threshing stone became a part of the Bethel College campus: "In spring 1903, C. H. Wedel, the first president of Bethel, reportedly set up a threshing stone in his front yard. The "Bethel College Monthly" of June 1903 was a bit bemused in its tone, 'Professor C. H. Wedel recently acquired an interesting relic in the shape of an old threshing stone, such as were formerly in

with the grain still in the heads was cut a few inches above the ground with sharp metal knives such as a small *hand sickle.* Larger knives, called *scythes* were used for faster reaping. But the largest cutting tool was the *cradle scythe*, which not only cut the straw, but with several strokes would fill the tines with a pile of wheat, and when dumped on the ground created a nice pile ready for the next men to bind into *sheaves* or bundles, tied with straw or later twine.

The cradle scythe was faster, but also a back-breaker, requiring operation by a strong man with great endurance.

The sheaves were then gathered into clusters and set upright into shocks to allow for further drying and ripening until ready for the threshing process.

The shocks were later transported, often with a *Leiterwagen* or ladder wagon pulled by draft animals, to a central location usually near a barn on

Threshing with a threshing stone in front of a typical Russian Mennonite house and barn - Drawing by John Klassen, ca. 1950. MLA

Threshing In Worms, Russia - Circa 1910. AHSGR

the farmstead. Here they were either piled into large stacks for later threshing or spread directly onto the *threshing floor*.

A circle of hard, flat ground was prepared as the threshing floor. This was often about a fifty-foot diameter circle, where the bare ground was prepared with water and compacting rollers (even using the threshing stone itself as a compactor). The floor would often be made even harder with straw and chaff embedded into the mud to create an almost adobe-like surface when dry. This created the ideal, smooth and hard surface for threshing and collecting the grain. The threshing floor was maintained from year to year and may likely have been used by multiple families.

The threshing process could take from days to months to accomplish. The sheaves could be thrown from the wagon onto the threshing floor or often re-

stacked into large hay stacks, to be threshed at a later date. The sheaves would be stacked to a tall height in such a way so as to deflect the water off to the side, keeping the majority of the straw dry until threshing was accomplished.

Writings talk of several methods of how the threshing stone was used, but typically the stalks were piled onto the threshing floor in a circle to a depth of 12-16 inches. Many writings talk of laying the heads all in one direction, and then being turned the other direction for additional rounds of threshing. However, the old photographs do not bear this out; the straw seems completely randomly scattered.

Threshing Stone with steel towing frame. MH&AM, Goessel

general use in Russia. This valuable relic of former agricultural methods now occupies a prominent place in Professor Wedel's yard, but there is little danger that anyone will carry it away.' The house and yard where Wedel displayed it sat where the Fine Arts Center is now."

Bethel Campus 1934 - MLA

John continues on to explain how it became the symbol of Bethel College, "On November 16, 1934, the threshing stone was adopted as the official symbol of Bethel College.

Presumably it was the decision of Edmund G. Kaufman. He had become president in 1932 and was energetically and vigorously molding the college. A brief note in 'The Collegian' reported in a brief ceremony, the two threshing stones, one of them the one that had been in C. H. Wedel's yard, were placed in front of the old Science Hall, on either side of the main door. Apparently this was envisioned as a new tradition for the college, an annual freshman initiation ceremony at the threshing stones.

Following is an excerpt from "The Story of Bethel College" by Peter J. Wedel and edited by Edmund G. Kaufman, in 1954. "In front of the en-

The stone was attached to a wooden or steel frame (several different styles were used), and the frame supported a solid steel axle, in most cases that went through a hole in the center of the stone. A single horse or team of horses would pull the threshing stone or stones around in a circle, starting at the outer edge and working their way towards the center.[3]

The horse pulling the threshing stone would often be controlled by a young boy. This was a rite of passage, allowing the young man to contribute real work to the threshing process.

The heavy stone, usually with 7

Wooden grain shovels were carved from a single piece of wood, in the 19th century farmers believed metal would bruise the seed. Author's wife Schmidt family display at the Goessel Mennonite Heritage and Agricultural Museum.

ridges, was designed so that when it rolled over the straw the ridge design would create a thumping action, that was strong enough to knock the grain loose from the head, but not so heavy as to crush the grain.

The straw was turned several times with wooden forks as additional rounds were made typically working from the outside edge in and then back out again. Eventually the straw was rolled, raked or tossed to the outer edges, being moved with wooden rakes. The straw was saved for winter bedding and fuel.

The remaining chaff and grain was swept to the center of the threshing floor, in preparation for *winnowing*. Winnowing is the process of separating the wheat from the chaff. Winnowing involved lofting the chaff and grain into the air with a *scoop shovel* and letting the wind blow the lighter particles, the *chaff*, to the side and allowing the heavier grain to drop to the floor. Of course this process only worked on days with adequate wind so some planning ahead was required.

Beautiful handmade wooden fork and wooden rake. Kauffman Museum

It was apparently believed that using a wooden shovel, carved from a single piece of wood, was preferred to a metal shovel because the wood would not bruise the grain.[4]

Additional cleaning after wind-winnowing could also be accomplished with the use of sifters. The mostly clean wheat and remaining chaff would be put into sifters and shaken to allow the grain to fall though the screen while the longer chaff particles would remain behind, providing cleaner wheat in the process.

Many farmers continued to do hand winnowing, but by the mid-1800s the winnowing machine had been developed. Hand-powered winnowing machines would be preferred for this process, being faster and cleaning better. Many of these machines were manufactured in the Ukraine by Mennonite agricultural industrialists.

The mechanical winnower works by feeding the grain and chaff into the hopper while the hand crank turns a radial fan blade, producing an updraft as the mixture passes through the machine. This was a great improvement in quality and labor savings over the old methods.

Grain Sifter or Winnowing Sieve - circa 1870 of wood and perforated cowhide. Kauffman Museum

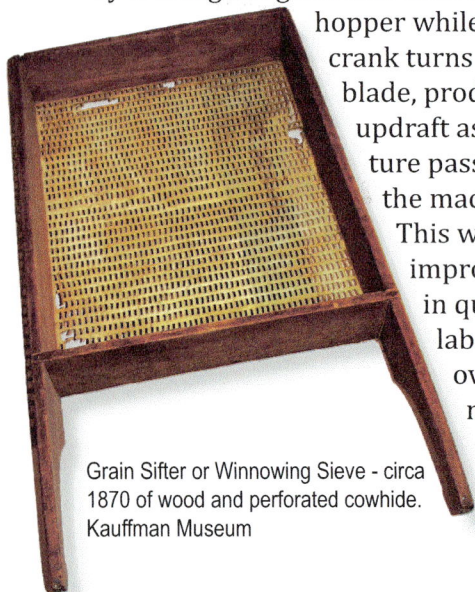

Round Sifter, circa 1870 of wood and wire - Author's wife Schmidt family display. MH&AM, Goessel

The clean grain would then be collected and stored, either in sacks or in bulk, later to be milled and used to make bread or sold at market, with some seed being saved and used for seed-wheat for the following year's crop.

[See Article 3 for more in-depth details about threshing history and Article 11 for first person accounts.]

Hand powered winnowing machine - Invented by Heinrich Sommerfeld - 1876. MH&AM, Goessel

trance on either side is placed a threshing stone, reminiscent of the primitive agricultural methods of our European forebears. The threshing stone was adopted as the official symbol of Bethel College on November 16, 1934. It is peculiarly fitting as such a symbol, not only for Bethel College but for the Mennonite church as a whole. The following pertinent points are gleaned from a description of the exercises which took place on the occasion referenced to above.

"Its appropriateness lies in the similarity of the laborious process involved in this method of separating the grain from the straw. The manner in which these threshing stones were used will be of interest to the reader, who knows only of modern methods of threshing...

"For Bethel College the stone appropriately symbolizes: (1) The Pioneering Spirit – Bethel College is the pioneer Mennonite institution of higher learning in America. (2) Simplicity of Life – Bethel College stresses the simple life. (3) Faith and Stability – Bethel College was built upon faith by its founders and has weathered the storms of more than sixty years. (4) Solidity of Character – Bethel College is striving to develop sound Christian character.

1936 photo taken by the Science Hall - MLA

"Both the stones…were made in this country and were actually used for threshing grain during the early pioneer years. One was made for H. Richert of the Alexanderwohl community. It passed from him into the possession of C. H. Wedel… When J. R. Thierstein purchased the property, the stone came into

The 400-800 pound limestone thresher was about 23-24 inches in diameter and about 29-30 inches long, with 7 teeth (that look similar to gear teeth) cut into the cylindrical surface. *[See Article 4 for more in-depth details about limestone and see Articles 5-6 for more in-depth details about the design and making.]*

The stones in the Ukraine were, in most cases cut by Mennonites, in the 1850s selling for as high as 18 Rubles. This industry kept as many as 14 stone cutters employed.[5]

The threshing stones were used in the Mennonite colonies from about 1840 to about 1905.[6] The stones became the less frequently used method as the mechanization of threshing became more common.

The Molotschna Mennonites had turned the land north of the *Sea of Azov* from a scarcely used region into a sea of waving wheat that helped supply the demands of the Russian and European markets. But the Mennonite population was growing and the farmable land was

Typical 750 pound limestone thresher with steel towing frame. Jerry and Leann Toews

virtually all developed so farming opportunities were becoming limited. Also Mennonites came to Russia for religious freedoms in addition to economic opportunities and now both their political autonomy and exemption from government service were becoming limited.

Russian collection of about 40 old threshing stones - Former Mennonite settlement of Gnadenfeld - 2010. Don Linscheid

In the latter half of the century the American West was being opened up to economic development. The *Atchison, Topeka and Santa Fe Rail Road* had obtained land and right-of-ways through the plains of Kansas. There was a need for people and commerce in order to make the railroad an economically viable venture.

The Kansas plains had about the same climate and soil conditions as in Southern Russia, and the plains were ideal for agricultural development. The AT&SF saw the opportunity to invite the talented farmers who had turned the Russian steppes into productive farm ground, and the Mennonites were looking for new economic opportunities and religious freedoms. So through brilliant and aggressive marketing tactics, the AT&SF offered thousands of acres of virgin farm ground on very good terms for them to relocate to Kansas, in McPherson, Harvey, Marion, and Reno counties. *[See Article 8 for more in-depth details about the railroad.]*

"WildWest" The Needs Immigrants

After much consideration and several scouting missions, hundreds of families traveled by ship to New York, then by rail to Topeka, Kansas, and ultimately in the fall of 1874 arrived in central Kansas.

They left most possessions behind, but brought with them, in large trunks and gunny sacks, the essential possessions that they needed to start a new life. They also brought with them the skill and work ethic that had served

1874 AT&SF Rail Road Co. Letterhead. MLA

them well in the past and hoped to establish new homes and excel in the breaking of the prairie soil. Times were hard, but most did succeed. *[See Article 9 for more in-depth details about picking Kansas.]*

In those trunks were some carefully selected hard Turkey Red winter wheat kernels that would be planted in the newly broken soil to be harvested the following summer.

Shortly after the Mennonites arrived in Kansas the stones were crafted. Mennonite individuals may have made their own or writings say that the threshing stones were ordered in mass from Florence cut by a stone mason as

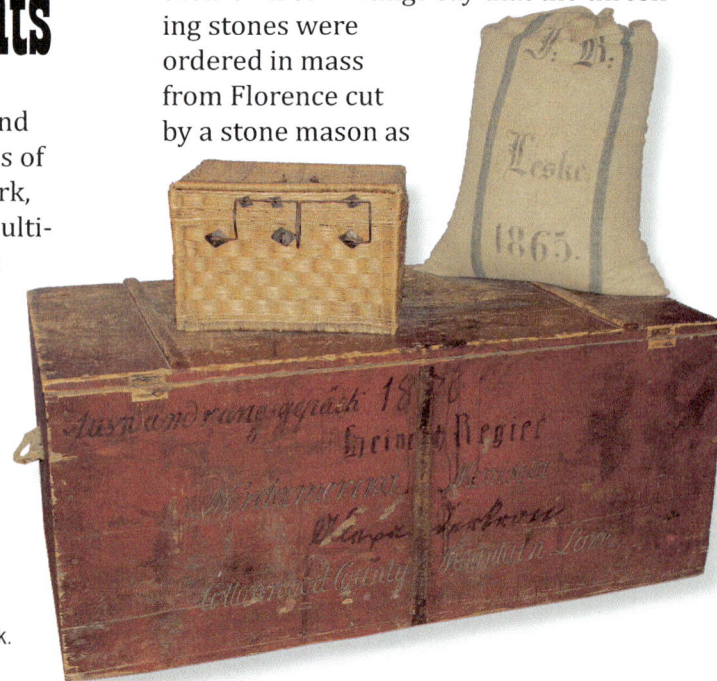

Travel trunk and grain sack. Kauffman Museum

his possession. It was donated by him to Bethel College. The history of the other stone is obscure. After passing through several different hands, it came into the possession of the R. A. Goerz family, who presented it to the Museum."

There are currently 4 original old stones on the Bethel Campus with two of them now located at Kauffman Museum. The stone now on Thresher Stadium is not original but a concrete reproduction. But which two are the original stones?

After about a year into my project I decided to study all the old photographs that included the threshing stones on the Bethel campus in order to determine where they had been and in what years. The earliest photographs that I have found are from 1934, showing two stones setting in front of the Science Hall. There are several photos that show these stones in this setting, and both stones are displayed with wooden towing frames in front of them.

Threshing Stone #4

One of these stones now identified as stone #4

is a square tooth design. Upon my review of the details of this stone, such as distinguishing fossil holes in the end, an enlarged axle hole on the other end, as well as a very distinctive chip on one tooth, I can see that this stone has been all over campus.

It is first seen as the stone in front of the Science Hall on the north side of the sidewalk. It can be seen in front of the old gym (Old Kauffman Museum building) in the 50s, later to be shown on a white rock circle in front of the Old Kauffman Museum entrance near the street in the 60s and later setting on a concrete pad in the 70s.

It should also be noted that in studying the old pictures that this stone and the other

1950s in front of old gymnasium. MLA

Topeka Train Station - 1867 - Similar to what the immigrants would have seen in 1874. Kansas State Historical Society - Photo Alexander Gardner

directed by Dietrich Gaeddert and Peter Balzer.[7]

There are many theories as to why the threshing stones have seven teeth, but the best guess is that through experimentation this configuration just worked the best for knocking the grain out of the head without causing damage to the grain. There is evidence of other sizes and shapes in other parts of the world, but this 7-tooth configuration seems unvaried in Kansas. (Many Mennonite writings talk of a conical shaped stone, but my research has not found this to be very common, based on the stones measured in Kansas. Drawings of threshing stones from other cultures do show a significant conical shape.) *[See Article 6 for more in-depth details on the design of the threshing stone and why 7 ridges.]*

In the first years not many bushels were

harvested and it took several years to establish a market for the new hard winter wheat, which was harder to mill. But within a few years it was the wheat of choice and Kansas was considered "The Bread Basket of the World." *[See Article 12 for more in-depth details about Turkey Red wheat.]*

The threshing stone had only a short life in Kansas. It is unknown how many were ever made, writings say as few as 28 and as many as 200, but we really don't know for sure. At that same time, the thresh-

1870s travel luggage. Kauffman Museum

Threshing stone with no axle hole. Jerry and Leann Toews

ing machine had already been developed and was starting to be marketed to those same productive farmers, who were always looking for a better way to improve production and productivity.

So many stones appear to have never been used and some were never even finished, evidenced by the fact that the axle hole was never drilled on many of the Kansas stones. Some threshing stones were also re-purposed; some had metal rings attached to be used as a place

A cold end to the threshing stone, re-purposed to be a salt lick in cattle corral. Wall family

to tie horse reins. Other threshing stones had one end hollowed out to be used as salt lick stands for cattle. But the threshing stone quickly became an outdated artifact, relegated to being lawn ornaments or just dumped into a ravine to minimize erosion.

My quest to find the stories of the threshing stone led me to many places and interesting discussions and revealed some surprising information.

Threshing stone re-purposed as a rein tie - Burrton, Kansas. Note many fossil holes in the limestone.

I have found almost 100 different stones in North America. All but 3 appear to have central Kansas connections. Most all have a Mennonite ethnic connection, but a few do not have any known Mennonite roots, and one appears to have Lutheran ethnic roots. Almost all are identical in size and design. All but two, found in North Dakota and Texas have 7 teeth.

one in front of the Science hall had a wooden towing frame both made identically. The distinguishing features of the frame

1960's in front of old Kauffman Museum. MLA

are the round timber side rails, with one wooden cross bar held in place by a mortis and tenon joint secured by nails. The axle was made of pipe with a pipe connector on each end to hold the axle in place. It appears that at one point this wooden frame was painted white. But in photos from the 50s I noticed that the wood had significantly rotted away. In later photographs from the 60s it has a new frame, almost identical except for the way the tenons were made, but it is clearly different. This style of towing frame is what is often depicted in Bethel graphics throughout the years.

The actual threshing stone itself most often depicted in Bethel graphics is this more common square tooth stone design. There are many examples of graphic images over the years at Bethel and virtually everyone has the square tooth design.

1970's photo...needs a cleaning. MLA

This graphically inspiring stone once heavily darkened with mold in the 70s is now clean and preserved in the permanent Kauffman Museum display. This stone with its distinctive chip is located indoors with the chip back up against the wall.

Square tooth design.
Author's collection

There are only two basic variants, that being the tooth design - either a square tooth or a round tooth.

All the stones with Kansas connections appear to be cut out of identical limestone, typical of limestone deposits in Marion and Chase Counties, but this finding required some interesting research.

Some writings spoke of 100 threshing stones that were made in Russia and shipped over for the Mennonites on the Red Star Line at the expense of the Atchison, Topeka and Santa Fe Railroad. *[See the end of Article 4 for all the details about this story.]*

The other surprising finding was the limited users of the threshing stones. All traceable Mennonite threshing stones appear to have been used only by Low German Mennonite groups that had immigrated from Prussia to Russia then to Kansas in 1874-1875.

I was surprised to find no threshing stones used among the Swiss Volhynian Mennonites, Whitewater Prussian Mennonites, Holdeman, Amish, Old Mennonites, or other ethnic communities outside of central Kansas.

Even though Mennonites also migrated further into many Western Kansas counties, I have not found any threshing stones in those communities. Also despite rumors of threshing stones in other counties such as Ellis, Russell, and Rush counties, no threshing stones have been located in areas colonized by the Volga Germans of Roman Catholic, Lutheran, and Baptist denominations.

I hope that this information does not, in any way, alienate individuals who have an emotional connection to the story, heritage, and symbolism of the threshing stone. I feel the threshing stone is a strong icon for many reasons and is worthy of the symbolism applied to it.

Bethel College threshing stone.

Cereal Grains

One cannot tell the complete story of the Threshing Stone without starting with the history of cereal grains. It has been said that "Bread is the staff of life" and bread is even used to symbolize the Body of Christ. Bread would not be possible without the ability to grow and process cereal grains.

It all starts with the need for food. Food is essential for human survival and varieties of plants and meats have been used for our sustenance since the beginning of humankind. We are hunters and gatherers; one of the things we learned to gather was grain. The development of grain agriculture in prehistoric times was certainly one of the great turning points in human history. There's glamour attached to hunting down wild animals, but from a hard-nosed, analytical standpoint, there is a more practical way to provide food.[1] The development of grain as food is one of the greatest stories of humankind and it has evolved throughout the history of civilization. The story of wheat is "The Story of Civilization." No one knows exactly when or where wheat was first planted but there is evidence that it has been around a long time.

Wheat's beginnings can be traced to a clan of wild grasses called *Triticeae*, the seeds of which had a flavor that was pleasing to primitive people. Triticeae included wheat, barley, rye, their wild relatives, and a number of important wild grasses. The Fertile Crescent, at the core of western Asia and northern Africa, is the center of origin and early diversification of this clan. Wild *Einkorn* and *Emmer*, which have been known for roughly 75,000 years, are credited as wheat's earliest ancestors. The ripple effect of these grains has been immense since wheat is the most widely produced and consumed cereal grain in the world.[2]

Evidence points to 10,000 to 12,000 years ago when wheat was first agriculturally grown in Mesopotamia and in the Tigris and Euphrates River valleys in the Middle East. Many anthropologists speculate that primitive people probably chewed the wheat kernel before they learned to pound it into flour and mix it with water to make porridge.[3]

The discovery that wheat could be selectively grown, harvested, and used to make food, as well as a food that could be stored for consumption at a later time along with the ability for the seeds to be saved for sowing the next season, was a significant node in the evolution of humankind.

Opposite page: Kansas wheat - 2012.

Tomb of Menna - c. 1422-1411 BCE. Wikimedia Commons

With the ability to cultivate food, rather than just following animals to hunt for food, groups of humans could remain in one place. This development in human skills allowed for the creation of communities, establishing commerce and community interdependency. Food production added to the development of commerce. As tribesmen traveled to distant places, the grain could be transported with them and used for trade.

The concept of private property likely developed when early humans built shelters to stay near their fields and claim their crops. Villages developed where land was plentiful as people joined together for mutual protection. No longer needing to wander, they spent more time thinking of ways to improve their crops, their shelter and their way of life.

It also allowed for individuals to specialize in trades such as farming, allowing for others to provide skilled services to the community, because now they could depend on the farmers to provide food throughout the year. With farming, now there was a need to understand the seasons and the timing of planting for maximum return on their seed investment. These cycles needed to be observed and understood, documented, and then passed on from generation to generation. This created a need for an understanding of time by studying the stars and the seasons, creating the need for calendars to track and document the understanding of optimal planting cycles.[4]

In order to trade ideas and understand each other they needed to develop the use of language. To aid in communication, symbols were developed which mark the beginning of writing. Even the first semblance of communal organization in the form of government was also related to wheat. Egypt was one of the world's early wheat lands; its network of irrigation canals required a grand plan with supervision and organization.[5]

These changes led to the building of towns and cities, the expansion of trade, and the development of the great civilizations of ancient Egypt, India, and Mesopotamia.[6] Farming changed civilization.

Stone #4, Kauffman Museum, NN

Threshing Stone #2

My connection to threshing stone #2, now in front of the Bethel College Administration Building, is that it was probably the first threshing stone that I and many other students ever saw. It was placed on its new limestone perch in front of the Ad Building in the summer of 2010 as a part of remaking the plaza area. I know this threshing stone was in front of the Ad Building in 1977 when I created a drawing for the centennial celebration at Hoffnungsau Mennonite Church. Featured prominently in the drawing is this threshing stone amid a collage of historic farming tools. I can tell it was this stone because of the unique round tooth design. This round tooth is a fairly rare shape, it has a full radius along the edge of each tooth and is easily identified from other stones on Bethel's campus. (See my drawing in Article 14.) Is this uniquely identifiable stone one of the two original in front of the Science Hall?

This stone, with its round teeth, has become part of a new tradition for several years old now, "the touching of the stone," where new students to Bethel file past the stone at the beginning of their first year at Bethel and then again as they graduate to "Touch the Stone" signifying their transition in life. (See image in Article 13.)

I assumed for a long time, that this stone was one of these two original campus stones. But in closer review of the photos in front of the Science hall, the shadows do not indicate a "round tooth" design.

The first Bethel stones were both square cut, so the stone now in front of the Ad Building is not from Wedel's yard. It does not show up in any photos until 1966, setting on white rock on top of the mound

Access to water and fertile ground was fundamental in the development of farming communities; dependency on rain alone was not enough. We find many of the early civilizations began along major river systems. For example, the Egyptians settled along the Nile River, the Harappa culture along the Indus, the Chinese Empire along the Huang River and the Mesopotamian countries along the Tigris and Euphrates rivers. The river systems provided these early civilizations with a consistent source of silt from the yearly floods and water for the crops.[7]

The oldest archaeological evidence for wheat cultivation is in Southwest Asia, known as the Fertile Crescent, and shows evidence that *Einkorn* wheat was harvested and domesticated on the fertile farmland.[8]

Around 8,000 years ago, a mutation or hybridization occurred within ancient wheat, resulting in a plant with seeds that were larger but could not sow themselves in the wind. While this plant could not have succeeded in the wild, it produced more food for humans. In cultivated fields this plant out-competed plants with smaller, self-sowing seeds and became

Clay engraved with pictographic writing from the Mesopotamic city of Kish (Iraq), dated from 3,500 BC. Contains pictographs of heads, feet, hands, numbers and threshing-boards. Department of Antiquities, Ashmolean Museum, Oxford. Wikimedia Commons

the primary ancestor of modern wheat breeds.[9]

So what happened genetically? All plants are identified by their chromosomes. Every variety of wheat grown today has arisen from wild, fourteen-chromosome wheat, undoubtedly Einkorn. Einkorn and fourteen-chromosome wild grass crosses, created twenty-eight-chro-

The Norek, a machine used by the modern Egyptians for threshing corn.

Illustration from the "Encyclopaedia Biblica" - 1903. Wikimedia Commons

mosome *(Tetraploid)* wheats. Only one twenty-eight-chromosome species can be found in nature: Wild Emmer *(T. dicoccoides)*. It grows in the region comprising northern Israel, west Jordan, Lebanon, and adjoining southern and southeastern Turkey, western Iran, northern Iraq, and northwestern Syria.

Emmer *(T. dicoccum)*, which closely resembles wild Emmer, is the oldest and was once the most widely cultivated twenty-eight-chromosome wheat. Well-preserved spikelets of Emmer have been found in Fifth Dynasty Egyptian tombs—the bread bakeries from that period in Egypt's history used Emmer flour.

"GENETICS"!

Varieties of wheat that have forty-two chromosomes are the most recently evolved and most-used types of wheat. All of these varieties have been cultivated by humans (as opposed to growing wild). They are hybrids of twenty-eight-chromosome wheats and wild fourteen-chromosome wheats or grasses. Modern bread wheat varieties have forty-two chromosomes and evolved from crosses between emmer and goat grass, which is the source of the unique glutenin genes that give bread dough the ability to form gluten.

The fact that prehistoric people accomplished selective breeding of wheat is a testament to their powers of observation and curiosity. Through their efforts

Emmer spikelets "TriticumTurgidum" - Collected in India, 1926./ Maintained by the National Small Grains Collection, USDA. Wikimedia Commons

and, much later, through the understanding of the laws of heredity by nineteenth-century Moravian monk Gregor Mendel, wheat began to diversify.[10]

The early gatherers were also the first millers; they selected grains that could be most easily released from their *glumes* or *husks* and prepared into food. People parched, simmered, and ground these grains into prepared flat cakes.[11]

The great books of religions talk of unleavened bread as a reliable food

Harvesting wheat with sickles - ca. 1310. Wikimedia Commons

On the new green - 1966. MLA

on the green.
In the 70s this stone was placed closer to the Ad Building. (You must look very close to see that this is the round toothed stone.)
I learned from Walden Duerksen that Waldo Voth had

This is the round tooth Bethel stone, even though in this picture it looks a bit like a square tooth stone. MLA

made the wooden frame for the threshing stone at Alexander-wohl Mennonite Church (stone #21), and since all 3 frames were made identically, I can only assume that he made the ones that were at Bethel too. These frames first show up in photos as they sat on either side of the sidewalk leading up to the Science Hall in about 1934.

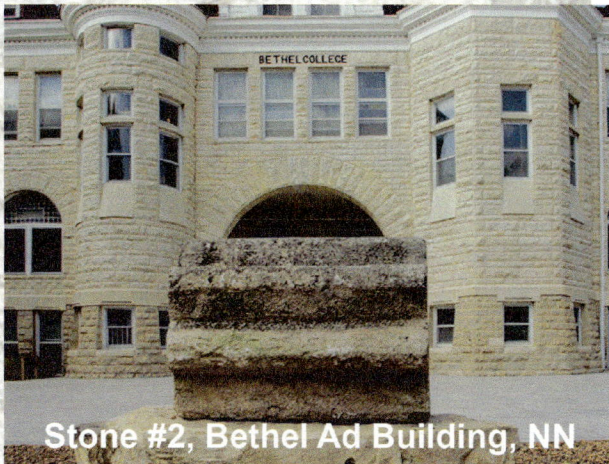
Stone #2, Bethel Ad Building, NN

The old wooden frame on this stone finally gave way to rot, captured in a photo as it sat on the sidewalk in front of the Ad Building the summer of 2011. This is the stone on the cover of this book.

The very next day the old wood was removed and in the fall of 2011 John Gaeddert

source. The numerous references to the growth of grain, which are found in the laws of Moses, indicate that it was expected that the Israelites would become an agricultural people after entering the land of Canaan, and that the cultivation of grain would become one of their chief industries.[12]

But other discoveries would come along in time. "It was the Egyptians who discovered how to make yeast-leavened breads between 2,000 and 3,000 B.C. They started fermenting a flour mixture by using wild yeasts present in the air. Eventually they added honey, salt, and flavorings such as poppy and sesame seeds. Even though the Egyptians and Romans made leavened bread, they did not understand that it was the yeast in the air that caused bread to rise."[13]

Since wheat is the only grain with sufficient gluten content to make a raised or leavened loaf of bread, wheat quickly became favored over other grains grown at the time, such as oats, millet, rice, and barley. The workers who built the pyramids in Egypt were paid in bread. Bread for the rich was made from wheat flour, bread for those who weren't wealthy was made from barley, and bread for the poor was made from sorghum.[14]

Now instead of just saving the grain for the future, bread allowed for a much more desirable food product which was delicious and more nutritious, plus it could be prepared in advance and be taken with them as they traveled about.

Abel Grimmer (1565–1630) - Der Sommer - 1607
Royal Museum of Fine Arts, Antwerp. Wikimedia Commons

Now food was portable.

The process of growing cereal grains has not changed much over thousands of years. The sequence of events is still the same today as it has always been; however the tools have changed.

These evolutions of wheat varieties and endless improvements are a testament to the powers of observation and curiosity by those individuals who worked those crops and continually selected the best of a strain which led to the varieties that were discovered and cultivated by the farmers on the steppes of the Ukraine, ultimately leading to the success of Turkey Red wheat that made its way to the United States.

There is no doubt that cereal grains, particularly wheat, played a significant part in development of the civilizations of the world.

The History of Threshing

Threshing is only one of the stages involved in getting the crop from the field to grain in the bin; it is simply the process of removing the grain from the head of a plant. There are numerous methods of threshing that have evolved over the last 10,000 years. In this article we will explore some of the processes to see how we end up with the threshing stone as it existed in the 1870s.

To remove the grain from the head of the plant, the plant is first harvested and allowed to completely ripen and dry. Then when the grain is most easily knocked loose from the hull, the grain is subjected to some form of violent shaking that releases the grain from the head. This process is called *threshing*. Threshing is the process that provides us with the grain that can be used for feed or processed into flour and used for human consumption.

No doubt those of us who grew up on wheat farms realize that threshing is easily accomplished by rubbing the head of grain in the palm of the hand and then blowing away the chaff, leaving only the grain. This process is still used by farmers in the fields today to quickly check the ripeness of the grain. If it is not ready for harvest it does not easily thresh from the head. To see if the grain is dry enough and ready to be harvested, experienced farmers can pop the grain into their mouth, and by chewing the grain, tell if it is ready for harvest. The hand threshing process was undoubtedly used before any other process to collect the grain. While on-the-go, people could grab some wild grain, thresh it in their hand, and chew the grain for nourishment. But hand threshing is in no way efficient enough to provide any usable quantity of grain. Humans eventually advanced to more effective means of threshing.

One faster method is to tear a fistful of wheat above the ground and beat the heads against a rock or large log. This ancient and most basic threshing process

Opposite Page: Threshing scene from the Egyptian Tomb of Menna, "Ancient Encyclopedia History" - Maler der Grabkammer des Menna. c. 1422-1411 BCE. Wikimedia Commons

knocks the grain and chaff from the heads where it can be collected from where it falls. This method can still be seen in parts of the world today where only minimal amounts of grain are collected for personal use, but this method is quite inefficient and very labor intensive, not suitable for large quantities.

The Threshing Floor - Greece - 2008. Wikimedia Commons - Stan Zurek

For most of history the threshing process required a large hard flat surface to more efficiently accomplish the threshing, i.e. Old English *threscan*, Dutch *dorsen* and German *dreschen*.[1] The threshing floor is nothing more than a large circular surface, ranging from 20 to 80 feet in diameter on which to perform the threshing process, allowing for clean separation from the grain and straw. Even in Biblical times the threshing floor was used for threshing and used as a symbolic metaphor in religious text. [For more on this see Article 13 on symbolism.]

The threshing floor was either hard-packed dirt or sometimes even covered with flagstones mortared together with sod, and often had a raised perimeter of stones to create a short wall that contained the grain on the threshing floor. This type of threshing floor was permanent and well-maintained to last from year to year. These threshing floors were likely used by an entire community throughout the threshing seasons. Many of these ancient floors still exist throughout the world.

The typical process in all of the threshing methods that use the threshing floor was to pile the straw, with the grain still in the head, onto the floor to a desired depth usually ranging from about 12 to 18 inches deep. Then one of various methods was applied to knock and shake the straw in order to cause the grain to separate and fall from the head. This was followed by turning over the straw several times with a type of hay fork and eventually removing the straw from the pile, leaving only the grain and chaff remaining on the threshing floor.

Illustration of winnowing from the "Encyclopedia Biblica" - 1903. Wikimedia Commons

Ad Building plaza stone, new frame - 2012

made a new hedge-wood towing frame for this stone in front of the Ad Building. (See picture at end of Article 13.) So which of the other two remaining campus stones is the other original?

Threshing Stone #3

This stone is also located on the Bethel Campus and is placed horizontally on a concrete pad at the side entrance to the Bookstore in the Schultz Student Center. At the time of my discovery of this stone I assumed this was the other stone that used to set in front of the Science Hall showing up in many photographs of the 1930s-50s era. However, this stone has a different past.

The finding of this stone came from a lead uncovered at the Goessel Country Threshing Days event where I had a booth set up to try to find information about threshing stones from individuals attending the event. Howard Goering told me that Fern Goering in North Newton had two threshing stones.

Photo Fern Bartel

I called Fern and made arrangements to visit her about these stones. She told me that at one time they did have two stones at their house but they had both been moved. One of the stones was donated to Bethel College in 2001, and had been placed in the raised flower bed by the main entrance to Schultz Student Center. Their

Treading Wheat - 2008 Tomás Jorquera Sepúlveda from Talca, Chile. Wikimedia Commons

Following threshing, the grain and chaff would be separated by a process known as *winnowing*. For most of history this was accomplished by throwing the grain and chaff into the air on a windy day. The wind would blow the lighter chaff off to the side and the heavier grain would fall down to the floor.

Earliest methods of threshing on the threshing floor included using the feet of humans and the hooves of animals to knock the grain loose. This process is referred to as *treading* or *trodding*, where animals such as oxen or horses were paraded around and around in a circular pattern on top of the straw. The straw was turned and eventually the grain fell to the threshing floor, where the straw could be removed and only the grain and chaff remained behind. This method can still be seen today in some countries where mechanized threshing is still not affordable.

Threshing with flails - Circa 1270. Wikimedia Commons

Early threshers may have chosen to beat the straw with a simple wooden stick. The short, club-like piece, called a *swiple,* struck the grain, and in time, pounding the grain from the head with sharp blows.[2] But at some point that method was greatly improved by adding a simple joint to the end of the stick. This was the invention of the *flagellum* [Latin] or, in English, *flail*.[3]

The flail is one of the most commonly used and long-lived methods of threshing grain around the world.

The flail is constructed with a long wooden handle held in both hands. At the lower end of the handle a flexible joint is used to allow a smaller wooden club to be "flailed" around in a circular pattern, creating much more frequent and more violent glancing blows on the grain to shock the grain from the head.

The proper use of a flail requires considerable practice and coordination. The ideal method is to establish a rhythmic cycle, allowing the club end of the flail to ricochet off the surface of the straw to allow it to continually be swung around and around without interruption to the flow. But this was still very labor intensive, tiring all but the strongest of operators very quickly. An output of 10 bushels a day was exceptional, seven was average. Some called flailing "the most degrading of occupations."[4]

Even today people around the world use this method to thresh grain where mechanical mechanisms are not affordable. Images of flails from around the world show a similarity of construction with slight regional differences in material and construction. The flail typically used by the Mennonites in the Ukraine and

Above - Modern Sherpa flail. Golden Colorado Restaurant

Below - Flail joint made of wood, steel and leather. Kauffman Museum

Flail - circa late 19th century. Kauffman Museum

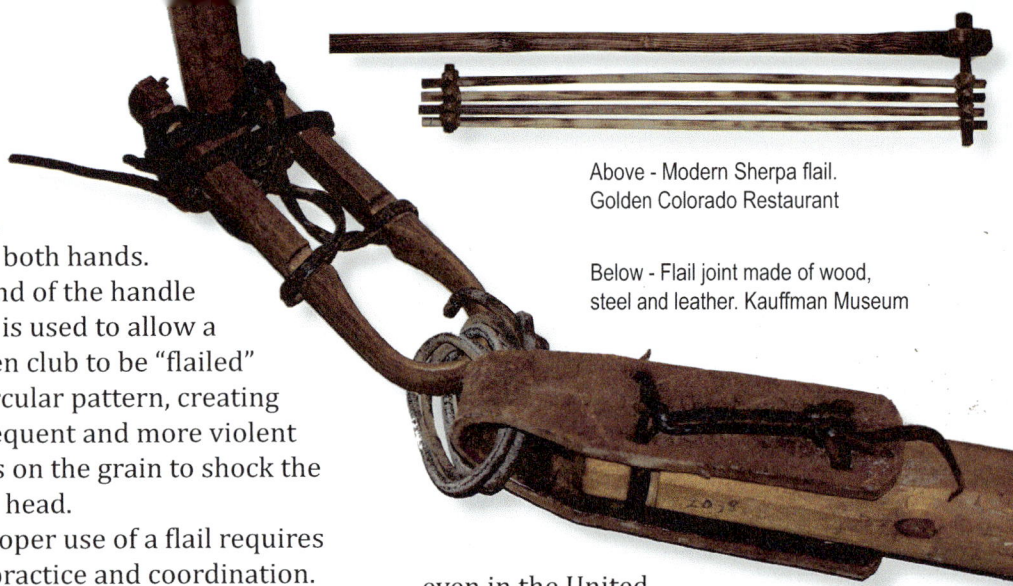

even in the United States is constructed of wood with a leather or a metal gimbal that allows for rotation of the club.

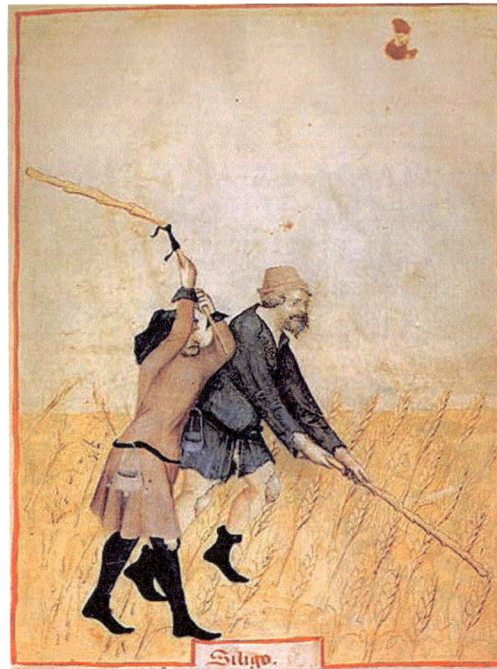

Peasants threshing wheat. Illustration from an edition of "Tacuinum Sanitatis" - 15th century. Wikimidea Commons

other stone had been moved back to her home farmstead.

This stone came into their possession when she and her husband purchased it from a man with five stones who lived west of Hillsboro but she could not recall his name.

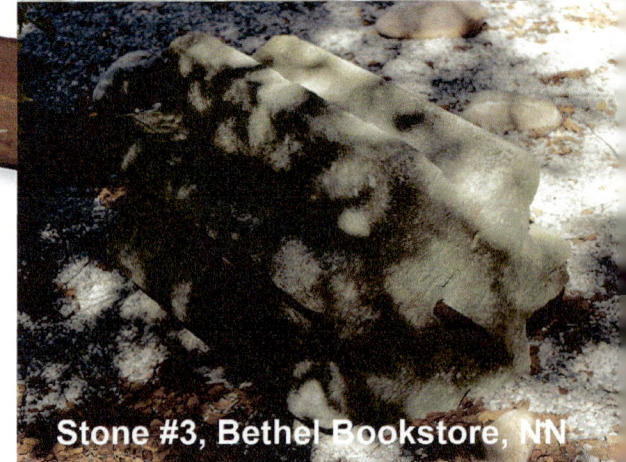

Stone #3, Bethel Bookstore, NN

I later went to the flower bed, which at the time was largely overgrown. I poked around and looked but saw no stone. For a while this caused me some concern, "where is it now?" I eventually confirmed with Bethel Maintenance Department that this stone had been moved from the flower bed to the concrete pad by the Thresher Bookstore entrance.

Threshing Stone #5

This stone is also a part of the Kauffman Museum collection and is located in the old Ratzlaff barn. In October 2010, I had a booth set up at the Buhler Frolic; I talked with John Ratzlaff who said they used to have two threshing stones on their farm, located at the in-

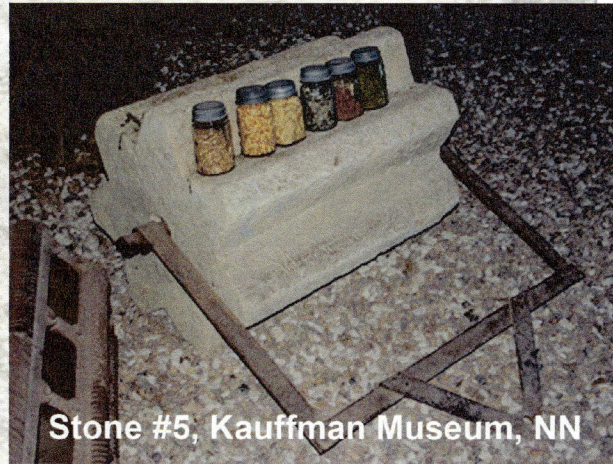

Stone #5, Kauffman Museum, NN

tersection of Dutch Avenue and the Bur-Mac road in western Harvey County. He said one of their stones got moved with the barn to Kauffman Museum. This is the one now in the barn. As of yet the other Ratzlaff stone has not been located.

The Romans show the earliest methods of mechanical advantages combined with the power of draft animals to greatly improve productivity.

"The *threshing sledge* or *trebulum* may possibly be the oldest farm implement in recorded history. Another clay tablet uncovered around 3000 BC has a primitive representation which archeologists have identified as a threshing sledge."[5]

Cylinder-seal impression from Arslantepe-Malatya (Turkey). 4th millennium B.C. It shows a ceremonial threshing with threshing-sledge. Wikimedia Commons - José-Manuel Benito - Public Domain by author Locutus Borg

The ancient Egyptians used wooden sleds or stone drags for threshing. They purposely roughened the underside of the sled to create the desired shearing and straw-chopping action. Around 720 BC Isaiah wrote: "Lo I will make you a new threshing sledge furnished with sharp teeth" Isaiah 41:15.[6]

Traditional threshing with a threshing-sledge of Middle Orient - 1884. Wikimedia Commons - Bishop John H. Vincent

Later, in Roman times, Marcus Varro (116-27 BC) recorded the following observations on threshing methods: "That threshing in his time was done on an open floor with either a *Tribulum* or a *plostellum poenicum.* A Tribulum was a weighted sledge with pieces of stone or iron embedded in the bottom to rub out the grain. A plostellum poenicum or *punic cart* was an axle, fitted with low wheels, upon which the driver could sit; it was used in regions along the Mediterranean. Sledges and rollers remain in use to modern times in the Middle East.[7]

Punic Cart - 1918 - Lekegian, G. - Wikimedia Commons

Threshing sledge - with stones in bottom to thresh the grain as it is dragged over the straw - 2007. Wikimedia Commons - Osmanyaylali

The basic sledge can still be seen in use today in many poorer parts of the world. Even though, it does not seem to be the ideal solution for effective threshing, it has a very long history of functionality.

Egypt Punic cart on threshing floor.
Courtesy image from LifeintheHolyLand.com

The Latin noun for threshing board is *tribulum*, derived from the verb *tribulare* that, literally, means to break something, to crush it. It has the same root etymology as tribulation - which is a tormentor or an adversary who persecutes someone.[8]

The punic cart is akin to the threshing stone in that the thresher rolled over the straw, but the cart was a more complex device with multiple axles and multiple disks of either wood or stone that threshed the straw.

Many places developed a rotary threshing device referred to by us today as the *threshing stone*. These devices can be found in almost all countries that have ever grown wheat. They are called by dif-

Punic cart detail - One of many designs that threshed the straw - Village Gakyh Doroud Faraman Blog. Wikimedia Commons

ferent names: In German it is *Dresch-Stein* or *Dreschwalze*, in Dutch it is *Dorsen Steen*, in Low German it is *Utfoasteen*, in Italian it is *Trebbiatura Pietra*, in Portugese it is *Debulha Pedra*, and in Russian it is *обмолот камень*.

Although the roots of the threshing stone evolved from the punic cart, early use of a threshing machine, that closely resembles the rolling threshing stone design, can be found in China.

A book on ancient farming methods of China shows an image of this

A Bethel Mystery? Where is the other square tooth stone?

So, if you have been following closely, the two original stones had square teeth. Of the 4 stones now on campus that we have identified, the one in front of the Ad Building has round teeth, so it's not original. The one by the bookstore has square teeth; but was donated to the college in 2001; the other in the Ratzlaff Barn also has square teeth but was donated to the Museum in 1984, and the one at Thresher Stadium is concrete. So where is the other original square toothed stone? I don't know; we have mystery. [See stone #22 for more].

Threshing Stone #6

This stone came to my attention when I first started this project. One of the first persons I contacted about my project was John Thiesen, Co-Director of Bethel Libraries and MLA archivist. I had attended his faculty lecture titled "What is a Thresher?" in 2008. I was already interested in the subject, so I was completely fascinated with all the images and history he presented, which convinced me that I needed to do this project.

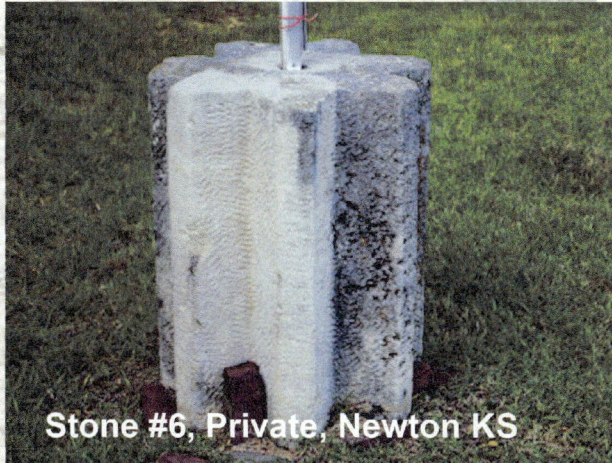

Stone #6, Private, Newton KS

Later when I discussed my project with him he said his father had one at his house in south Newton. His father had told him that his grandfather Henry Thiesen had several threshing stones. The one they have is the last one of those, but he didn't know what happened to the others. John's cousin had told him that there was one on the farmstead where another cousin now lives. She also thought there may still be a broken-up one or more somewhere on the farm.

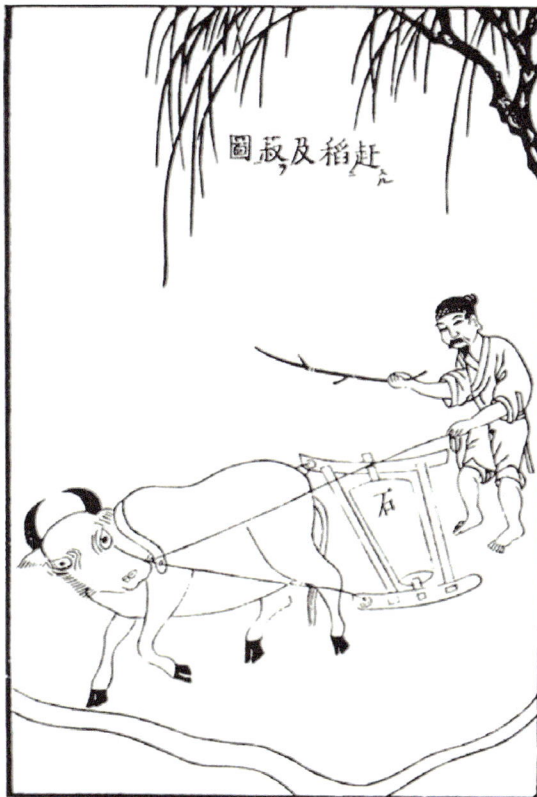

Illustration of threshing stone published in 1637, in the "Tian-gong Kaiwu Encyclopedia" written by Song Yingxing.
It can be found in E-tu Zen Sun and Shiou-chuan Sun's translation - Source Pennsylvania State University Press, 1966

Typical smooth surface threshing stone in Western China - Photo 1976.
Courtesy Robert Kreider

conical shaped stone being pulled around by an ox. This drawing shows an extreme conical shape allowing for turning in a smaller radius. But typical of many Asian stones found, the stone does not have ridges cut into it.

Threshing stones from around the world however typically do include ridges cut into the cylindrical surface--most often 7, but as many as as 60, as few as four, and in a few cases just a flat 6-sided hexagonal shape.

It is unclear exactly when the Mennonites started to use the threshing stone. The concept of a threshing stone certainly preceded the Mennonite farmers, but many writings give credit to the Mennonites for maximizing the development of the threshing stone. The use of the threshing stone does not show up in Mennonite records until about 1840, when wheat became a more prominent crop for Mennonites farming in south Russia. Some early threshing stones were reportedly made of large wooden logs, but soon the more common stone design was used exclusively.

The Museum of Archaeology and Evolutionary Ecology in Volgograd, Russia, has collected information on the Mennonite artifact. "Place of Discovery: Kamyshin - Age: 19th century - Weight: 100 kg - A step forward in the 1840s was the threshing stone that was pulled by horses.

Mennonite Threshing Stone. Museum of Archaeology and Evolutionary Ecology in Volgograd Russia. http://museum.vgi.volsu.ru

The horses walked slowly around the circle, the stone knocked the grain loose which then was collected in the middle. This method, which was widely used in the 60s, was continued in some of the Crimean colonies until the end of the 19th century."[9]

The method of using the threshing stone is documented in other parts of this book, but the basic concept was to roll the stone over the straw which was laid on the threshing floor, draft animals pull one or more threshing stones over the straw as it was repeatedly turned. The steady thumping of the heavy stone knocked the grain loose from the head.

Eventually the straw was removed, leaving the grain and chaff on the threshing floor. This method, though crude, was still considerably more efficient than most any other method known at the time.

China Threshing Stone with many small ribs. Maurine (Voth) Regehr

I have found evidence in photos and writings of threshing stone use throughout all areas where wheat has been grown--across Europe, the Middle East, India, Asia, Northern Africa, Australia and eventually North America.

Already mentioned was the evidence of use in ancient China. Mennonite missionaries that were located in north central China in the village of Daming recalled seeing threshing stones being used in that village in the 30s. A photo from a return trip to China by Maurine (Voth) Rehehr in 1987 shows at least one of those stones still remain. I have discovered over a hundred photos of remnants of threshing stones still located in the Ukraine, where they were apparently a very popular threshing tool.

Even in the U. S., George Washington wanted to improve the treading method. He is credited with developing a unique threshing barn. "The barn measured 52 feet in diameter with a 28-foot central octagonal section (used for storing unthreshed wheat). A 12-foot-wide oak threshing lane encircled the center section. Horses would run around and around within the lane, treading the grain out of the wheat. A farm worker was present to make sure that the horses did not stop running, because horses do not urinate or

Barb, John, and Andrew Thiesen...stone moved.

In 2011 I helped John move the threshing stone from his father's place to his place. In the process we weighed the threshing stone and found it to weigh 820 pounds.

Threshing Stone #7

This stone came to my attention while doing research at MLA. A college employee came into the library and stopped to chat. After explaining my project she said her mother-in-law had one at her place. There was some concern about her being comfortable with giving out information, but after a long time, I finally contacted her and she let me come over and take some pictures and listen to stories about the history of this stone. It came from her side of the family in the Goessel area. Her father had put a lamp on it for display in their yard. She remembers having picnics on top of the stone when she was young.

Stone #7, Private, Rural Marion

When she and her husband cleaned up their family farm they moved it to their home in town, and they moved it again when they located to her current residence where it now sets on end and is still used as a yard lamp base.

So Unobservant!

Even though I grew up in Buhler and went to Buhler High School I did not recall that there were 4 original stones on

defecate while they are running. Washington designed the flooring for the barn's treading level so that there were 1 1/2 inch gaps between the floorboards. As the horses treaded out the grain from the straw, the grain fell between the gaps to the first floor, where it was gathered up and stored until being taken to the grist-mill to be ground into flour."[10]

The 1811 *Encyclopaedia Britannica* tells of a tapering roller fastened to an upright shaft in the center of a threshing floor and pulled by oxen that was still in vogue in Italy, and suggested that it probably descended from the Roman tribulum, or from the roller sledge. The lumbered threshing roller was illustrated by J.C. Loudon in 1831. In an Australian newspaper article, it was noted that the carved beaters (fluting) are similar in section to the longitudinal grooves of Ukrainian-American stone Mennonite Rollers.[11]

Lumbered Roller Illustration. Loudon, "Encyclopaedia of Agriculture" - 1831, p. 49.

South Australian farmers were advised against using the parallel tillage roller for threshing, precisely because it pushed the crop up into a heap rather than rolling over it.[12]

They developed long tapered threshing machines, either from single

Various Australian threshers made of wood with taper. Source www.sahistorians.org.au

pieces of wood or assembled from multiple wooden logs. These would be attached to a single center post while being drawn by the horse who walked outside of the threshing area.[13]

Another tapered wooden device, know as *Dreschblock*, was often used in the Netherlands and Germany.

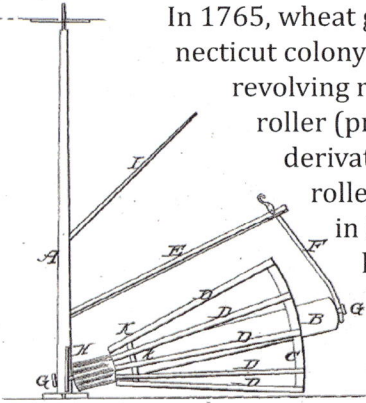

Hand thresher invented by George Anstine in 1843. United States Patent and Trademark Office.

In 1765, wheat grain in the Connecticut colony used a "large revolving ribbed cone" roller (probably of Dutch derivation). The conical roller was introduced in Maryland, in the later eighteenth century, and persisted beyond the introduction of efficient threshing machines. Such rollers were used in the wheat-growing areas in central New York state, and in parts of Maryland, and in Delaware.[14]

Michael Menzies, a Scot, invented one of the first mechanical threshers in 1732. It consisted of flails attached to a hydraulically operated wheel and delivered 1,320 strokes per minute.[15]

During the 18th century, many attempts were made to improve the threshing process, including combining threshing and winnowing into one process, but I have found it impossible to determine who was the first inventor of many of these machines. The history is quite unclear.

Another Scot, Andrew Meikle (1719-1811), invented a successful thresher in 1786 that consisted of a rotating drum with four vertical blades and a stationary shield through which the grain was fed. A fan blew the chaff away from the heavier kernels of grain, the process called winnowing.[16]

Around 1816, Robert McCormick, father of Cyrus McCormick, inventor of the reaper, experimented with threshers and other farm machines, but was not successful. In 1828, Samuel Lane of Maine invented a portable thresher that could also harvest-the first *combine* harvester.[17]

In 1834, John Avery Pitts and Hiram Abial Pitts devised significant improvements to a machine that automatically threshed and separated grain from chaff, freeing farmers from a slow and laborious process.[18]

"The first *combine* is often credited to Hiram Moore in 1836. Change wasn't accepted so it would have to wait till the end of the century for its turn. After time it did become the dominant harvesting method. It revolutionized the way the world ran. It was successful because it made farming safer, more profit-

Portable "Ground Hog" Threshing Machine - circa 1850. MH&AM, Goessel

public display in town. When I started this project in addition to asking people for leads I would just drive around with my eyes open and see what I could find. To my surprise I found 5 on display in town. Oh, the one on the south end of Main is a full-sized concrete reproduction made in the 1970s.

Threshing Stones 8-9

These stones have been in the front of Buhler High School for many years. They show up in many "Crusader" yearbook photos. My mother told me that when she was student at Buhler in the early 30s, one day she felt ill and one of her teachers took her outside to get some fresh air while sitting on one of the stones. These two stones now set upright on either side of the BHS sign in front of

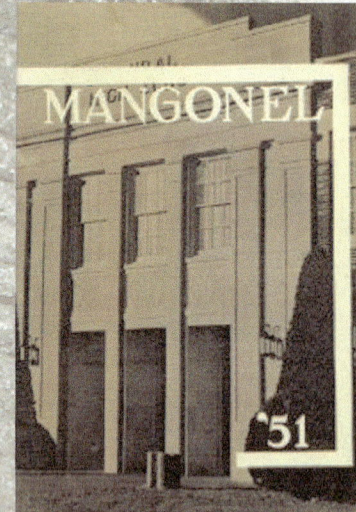

Stone #8, Buhler High School

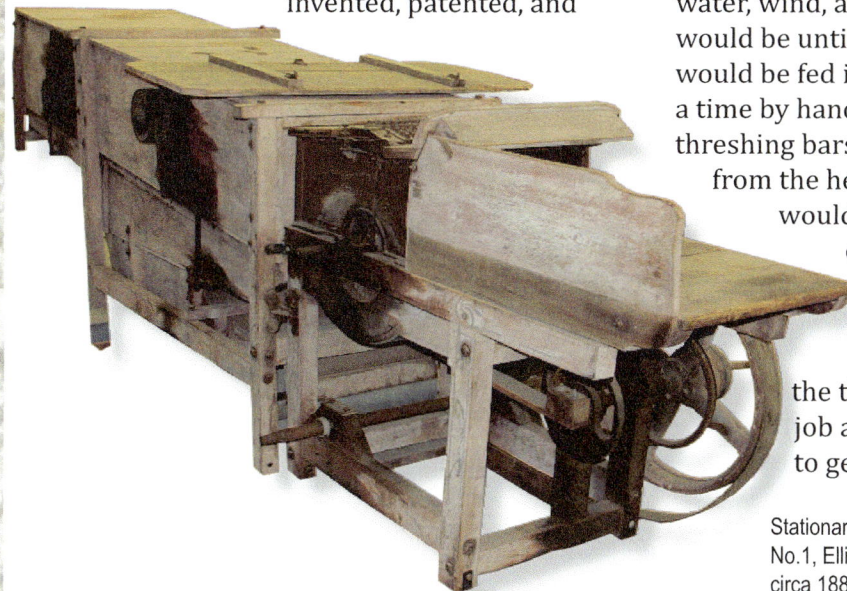

the main entrance along Main Street. But one story has it that while the last building addition was being added to the school the stones were thought to be rubble and were taken to the landfill. When someone discovered this mistake they had the construction company go back and retrieve them from being lost forever.

The origin of these stones is not known, but they are a long standing tradition of the school. One former student confessed to me that they had a plan to roll the stones into the entryway as a prank, but the plan was quickly abandoned when they found the stones just too heavy to move.

able, and brought food to many. But through the 1800s, the header and the thresher were king. The header was pushed through the field by six horses from the back of the machine. Pushing from the back reduced crop trampling. On the front edge of the header was a row of sharp teeth called sickles."[19]

Evidence of Mennonite inventions in the threshing machine also are known. "Many Mennonites were clock makers, but eventually the market became saturated, and modernity set in, clock makers re-invented themselves, using their mechanical skills to produce threshing machines, horse rakes, and reapers."[20] Additionally, harrows, wagons, cultivators, drills, winnowers, corn shredders, chaff cutters, weeders, fanning mills, and oil engines were invented, patented, and produced by Mennonite manufacturers, as well as by non-Mennonite Germans in Russia.[21] The annual output of Lepp and Wallman of Schönwiese, Ukraine in 1911 was 3,000 threshing machines.[22]

The portable, wooden-sided threshing machine could be moved from place to place, being powered by horses, water, wind, and later steam. The bundle would be untied, the straw and heads would be fed into the thresher one at a time by hand, being grabbed by the threshing bars that beat the grain loose from the head and straw. The straw would be transferred out the other end and a blower would winnow the chaff away from the grain.

Feeding the bundles into the thresher was a dangerous job and required great care not to get a hand grabbed by the

Stationary Threshing Machine - Champion No.1, Ellis-Keystone Agricultural Works - circa 1888. MH&AM, Goessel

The horse engine provides horsepower for threshing. MLA

machine.

In later years the threshing machine got bigger and less portable, often attached to the side of the barn or sometimes even inside the barn or under a lean-to. Power to run the machine was often horsepower collected from horses connected to a merry-go-round device called a *horse engine*, where the horses went round and round while connected to arms attached to a central hub that transferred power to the machine via a drive-shaft device.

Soon larger and portable threshing machines were developed and eventually powered by huge steam engines. These new threshing machines could work fast, were safer to use with automatic feeding belts,

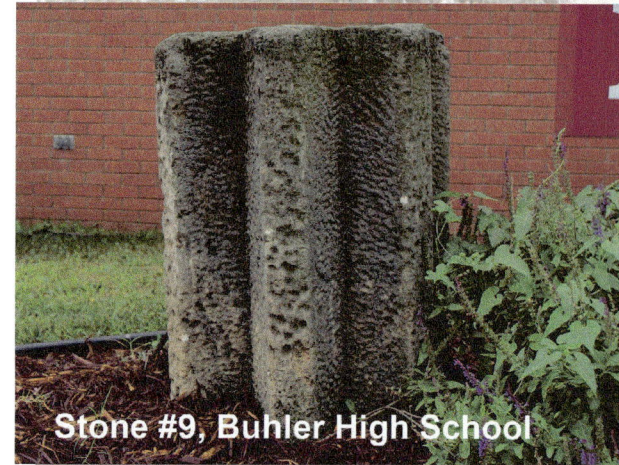

and had greater capacity.

This book focuses on the period up to the mid 1870s but it is readily apparent that the development of new technology has never stopped evolving. For most of history the process of harvesting grain remained a multi-step process of reaping, bundling, shocking, hauling, threshing the grain, winnowing, and collecting the grain--multiple steps that could often take months to complete.

The threshing machine is often considered one of the greatest inventions for reducing the manpower needed to feed the masses.

Threshing machines were readily available in North America when the Mennonites immigrated in 1874, but the implement dealers were not set up in central Kansas. However, without a doubt, it was not long before they realized the market potential and filled the need.

In the early years of threshing, machines were purchased by individuals who would gather together threshing crews to operate the machines. These threshing teams would go farm to farm

Threshing Machine - New Peerless, 1904. Goessel MH&AM

Stone #9, Buhler High School

The north stone of this set is the first stone I discovered without an axle hole. The lack of an axle hole clearly indicates that there is no way it could have ever been used for threshing.

Threshing Stones 10-11

These stones are located along Dutch Avenue at the north end of Buhler, Kansas, on either side of the historical monument "Our Mennonite

Stone #10, Buhler Monument

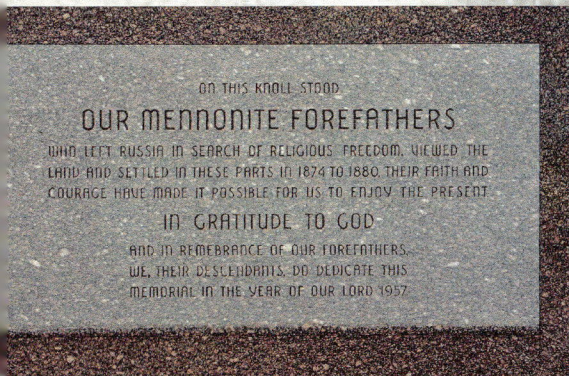

On this knoll stood
OUR MENNONITE FOREFATHERS
who left Russia in search of religious freedom, viewed the land and settled in these parts in 1874 to 1880. Their faith and courage have made it possible for us to enjoy the present

IN GRATITUDE TO GOD
and in remembrance of our forefathers, we, their descendants do dedicate this memorial in the year of our Lord 1957

Forefathers," erected in 1957.
They are cemented into the pad that they set on. Unfortunately after contacting many people, no one knows the origins of these stones.

Stone #11, Buhler Monument

Threshing Stone 12-13
While driving around I found these two stones in Inman. They have been there for years, I just never noticed. These stones

Gleaner, first self propelled combine. AGCO, Hesston

and thresh the grain for a fee. This process was quickly adopted by all farmers and made the threshing stone all but obsolete very quickly.

The pull-type combine, first pulled by horses and later with a tractor, was another major advancement, "combining" all the steps of harvest into one operation, all of which could be accomplished by one man.

The invention of the self-propelled combine comes from two brothers in Kansas in 1923, the Baldwin brothers. Their Gleaner Manufacturing Company patented a self-propelled harvester which included several other modern improvements in grain handling.[23]

The combine has continued to evolve. In the 1960s a 12' long header was typical; now they are up to 40' in length. The process of threshing grain has evolved to the even more efficient *rotary* threshing method. Many of the large combines today with their incredibly long headers are powered by a 400 HP Turbo Diesel engine and can collect up to almost 400 bushels of grain before needing to unload,

being able to unload that grain in only 1-1/2 minutes while "on the run."

Interestingly many of the modern monster combines are currently made right in Hesston, Kansas, at a manufacturing company called *AGCO*. They currently make combines for many brand names sold around the world, including *Massy-Ferguson, Gleaner, Challenger (Caterpillar), Fendt,* and *Valtra.* The threshing stone used to cost about $2.00 in 1874 and a modern combine and header may cost upwards of $500,000 today.

There are still threshing crews that follow the progress of harvest as it moves up the plains from Texas going north all the way into Canada. Many of these crews are based in central Kansas, in the same communities that first used the threshing stones in 1874--in Reno, McPherson, Harvey, and Marion counties.

No doubt change is inevitable as humans progress, and the threshing stone is but part of this evolution.

2012 combine, Kansas harvest - Greg Goering in McPherson County adjacent to Hoffnungsau Mennonite Church.

The Stone Geology

The threshing stone is all stone. It is a simple but archaic tool, made from a single chunk of rock. Since Roman times the threshing stone has been used to thresh grain, and was made from the most basic of materials...rock.

— "Surface" layer.

— "Fort Riley" layer from which the threshing stones were made. Fairly hard, very small fusulinid fossils, thick layer.

— Thin "Oketo Shale" layer between the larger limestone strata, broken into smaller blocks, calcareous, and contains marine fossils.

— Older "Florence" layer where chert is abundant and stone is highly fossiliferous.

Opposite Page: Florence, Kansas limestone quarry - 2012. Same limestone bedrock as in 1874. The boulders in the foreground would be large enough to make a threshing stone.

Rock has been used since the beginning of humankind; rocks, bones, sticks, and skins have been modified to make some of the most elementary of tools.

Rock has the constant characteristic of high density, which makes it heavy. It is available in virtually all parts of the world and can be formed to function for a variety of tasks as tools for pounding, crushing, grinding, cutting, and building. Each type of rock has different characteristics that yield to the manipulation of humans for the ideal fabrication of tools for specific tasks.

The threshing stone is one of these unique fabrications of humankind, made for the specific purpose of threshing grain. As discussed in other parts of this book, the threshing stone had evolved into a relatively consistent carved chunk of rock. That rock is about 300 to 800 pounds, cylindrical in shape, often with 7 ridges to pound at the heads of grain, with an axle attached through the center for the stone to revolve around.

The rock is always quarried from the locally available stone near the areas in which they are used. The requirements of the stone are that it be able to be carved into the desired shape, be the right density, and yet have the durability to withstand the multiple rounds of pounding out the grain.

The Lutherans also used the stone located on the west bank of the *Volga River*: "Two mountains of which the largest Urakov mound...found conchidal stone, very strong, which is used for the threshing stone and buildings."[1] The Mennonites were using threshing stones in the Ukraine from about 1840 to 1900 where the stone used was taken from the local quarries. "In 1855 the Mennonites, on the meadow side, knowing that there is a suitable stone, ordered (by model) threshing stones, and from that time began to be engaged constantly in making them."[2]

Of the stones I have found in North America, two are located in *Steinbach, Manitoba*, where, according to Roland Sawatzky, Senior Curator at Mennonite Heritage Village, these stones were made locally for public demonstration purposes. He can tell that they were quarried out of local *Tyndal Limestone*, because the grayish blotches within the tan limestone are characteristic of the limestone in that area.[3] One other stone found was made in South Dakota, from stone typical to that region.

The discovery that the majority of threshing stones used in North America are located in central Kansas directs us to the quarries in Marion and Chase counties. At this point I believe that all the threshing stones found in the central states have roots to the limestone found in this area. But this statement is in contrast to writings from several sources, saying that 100 were shipped to Kansas by the Red Star Line at the expense of the Atchison, Topeka and Santa Fe. *[See the end of this article for the full story.]*

So how did limestone get here and why was it chosen for the threshing stone?

We go way back in time to early Kansas lands or, actually, oceans. Evidence points to the fact that these oceans alternately covered Kansas, retreated, and then covered the land again. It was during a period from 570 to 230 million years ago when these oceans were coming and going, a period the geologists call the Paleozoic era, that the depositional environment of the seas changed slightly.[4]

"Limestone is a sedimentary rock composed mostly of calcite (calcium carbonate, $CaCO_3$). It is formed by organic means—that is, from the remains of animals or plants—or by chemical deposition. Many animals and plants (such as oysters, corals, some sponges, sea urchins, plankton, and algae) take calcium carbonate out of the water and secrete it to form shells or skeletons. As these organisms

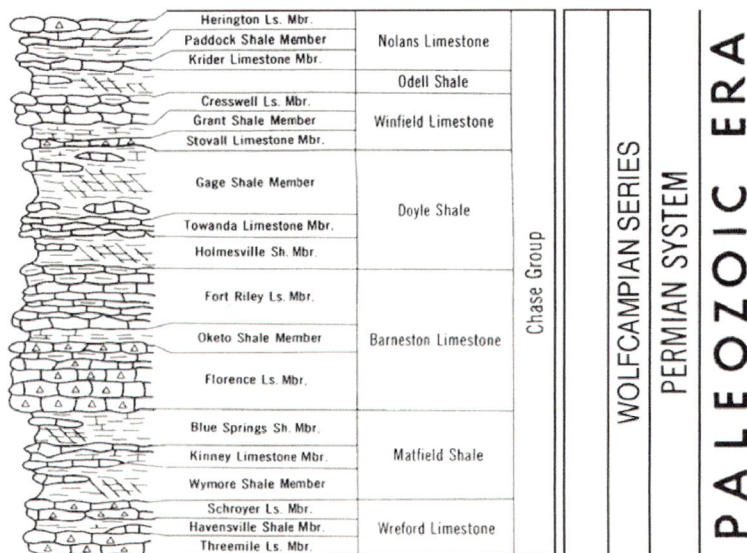

Stratigraphic Succession in Kansas. State Geological Survey of Kansas - Bulletin 189 -1968 - Zeller.

die, they drop to the bottom of the ocean. Over time, the organic parts decay and the calcium carbonate accumulates to form limestone."[5]

The *Paleozoic Era* laid down rock in central Kansas in the *Permian System* in the layers referred to as *Barnstone Lime-*

are located on Main Street in Inman, Kansas, in front of the Inman Museum. Nothing is known of the history of these stones.

A sign located next to the stones says: "Farmers in south Russia used threshing stones to thresh newly harvested wheat. The cut wheat was arranged in a circular pattern over a threshing floor. The seven-sided, cone-shaped wheels weighed from 360-720 pounds and were pulled by a team of horses in precise rotation over the wheat.

Stone #12, Inman Museum

As the stone knocked the wheat loose from the straw, laborers turned the straw and cleared the threshing floor for the next load of wheat. Mennonite im-

Stone #13, Inman Museum

Fusulinid fossils from the Topeka Limestone - Wilson 44691. Wikimedia Commons

migrants of the 1870s brought this technology to Kansas. Over time, they transitioned to the use of faster and more efficient mechanized threshing machines. Approximately 28 of these stones remain in the central Kansas area. They stand as a symbol of the region's rich agricultural heritage." Only "28" Huh, I think there are more!

Santa Fe Days

While spending a day at Inman Santa Fe Days with a table set up to try to find threshing stones, I learned that at one time these two stones had been located in front of the shelter house in the Inman park, where many Inman people will remember them. Later they were moved in front of the Inman Museum.

stone, in the stratigraphic units that crop out in Marion County, Kansas. This era can be divided into three segments. Lowest and oldest is the *Florence Limestone*, gray to tan thick-bedded fossiliferous limestone with many beds and nodules of chert and a thin shale parting. The thin layer above this is *Oketa Shale*, gray calcareous fossiliferous silty shale, used for fill material. The top layer is *Fort Riley Limestone*, which is gray to tan fairly hard fossiliferous limestone with thin shale parting with thick massive beds called *Rim-Rock*.[6]

As the limestone was formed, the stone often incorporated parts of animals that populated the ancient oceans. We call these remains fossils. *Cretaceous* and *Paleozoic* limestones often contain pieces of *Crinoid* which are extinct invertebrate animals that resembled a tulip and were nicknamed *Sea Lilies*.

Another common fossil found in limestone is a *Fusulinid*, the fossilized form of these one-celled animals resembles – appropriately enough for Kansas – a tiny grain of wheat.

Limestone, often with its trapped fossils, remained undisturbed through the ages, until Kansas pioneers began to use it for construction."[7]

The stone has a great many different physical properties. Hardness is determined by the degree of consolidation as well as by the actual hardness of component minerals, but even the densest can easily be scratched with a knife.

The color effects are caused mainly by impurities while the texture may be *amorphous semi-crystalline* or *crystalline*. The limestone having a higher degree of porosity may weigh as little as 110 pounds per cubic foot, and the more compact varieties may weigh as much as 170 pounds per cubic foot. The dense, highly consolidated forms are preferred for most uses.[8]

Chert and *Flint* – One of the naturally occurring forms of silica, SiO2, is dense and extremely fine-grained. Generally, light color materials are called chert while the dark (grayish) form is termed flint.[9]

The white limestone extends over the greatest area, and is more nearly uniform in leading characteristics than any other material in Kansas. This great belt of choice building material enters the state from Nebraska and reaches south into Marion and Chase counties. As the limestone erodes, angular fragments of flint accumulate at the surface, giving the Flint Hills their name.[10]

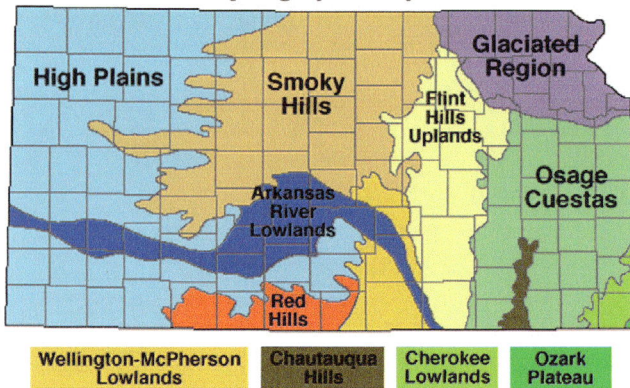

Generalized Physiographic Map of Kansas

Kansas State Geological Survey of Kansas.

In these counties, limestone reaches its greatest perfection, being quite white, and almost entirely free of the porous condition referred to above. The thickness of the ledges is from three to seven feet in layers or courses of from eight inches to five feet, furnishing material of all dimensions for any required use.[11]

A portion of the stone within the range referred to above, resembles in texture the celebrated soft stones of Europe. It can be sawed with a common wood-cutting instrument, can be bladed to a surface with a common wood-working tool, and yet, properly handled, it will carry a full load in the largest structures.[12]

The surface hardens instead of wearing with exposure, so that a sufficient length of time would render the exposed surfaces as hard as glass. The drying limestone brings calcite to the surface and creates a case hardening effect.[13] The evidence of long-standing structures emphasizes one of the major advantages of using Kansas limestone, its durability.[14]

During the expansion into the west, the discovery of the high quality limestone in the Marion and Chase counties inspired entrepreneurs to quarry and market this fine limestone. Initially the quarries were used for local railroad bed construction and to this day fine bridges, buildings and homes display beautifully the qualities of the limestone as well as the skills of the limestone stone cutters.

The *Topeka Commonwealth* newspaper of August 10, 1871 stated, "To the Editor of The Commonwealth. Knowing your desire to keep the readers of the Commonwealth posted on all matters pertaining to the rapid growth and development of Kansas, I take the liberty of sending you a few items concerning the

Threshing Stone 14

I first learned of this stone from an old friend of mine, Orlin Martens, who told me that his family had a threshing stone. I soon met with the owner Wilma (Friesen) Martens, and learned the story.

The stone comes from the farm where she grew up 1 mile east of Goessel. Her parents were Daniel and Suzie Friesen. They moved the stone

Stone #14, Private, Inman

to Inman, Kansas, where it was displayed in their yard for many years. It has since been moved to another location in Inman. This stone is in excellent condition and is displayed on its side atop a concrete slab.

Still Unobservant!

Even though my mother and father in-law lived just yards away from the three stones at the Goessel museum, I did not remember them being there, but I was quite sure that they would have some, and they did.

Threshing Stone #15

This stone sets on a concrete pad outside the Goessel Heritage Museum immigrant house. It has a horizontal crack all the way through it that has been glued back together. It was donated to the museum by A. Adolph & Velda Duerksen.

Stone #15, Goessel Museum A

flourishing condition of Florence and its surroundings. The town is now about four months old, and contains a population of some five hundred people. It stands on the west bank of the Cottonwood River, about forty-five miles west of Emporia, and is surrounded by a rich agricultural and stock growing country, abundantly supplied with the best quality of water, stone gypsum, and a very fair quality of timber. The stone is a fine texture of pure white lime, soft and easily worked. Inexhaustible quarries adjoin the town on the west, which are being opened by Mr. Horner, of Topeka, who ships by rail both east and west."[15]

The stone quarrying business continued to do well for many years after its beginnings in the 1870s. One of the first rock quarries was owned by A. F. Horner of Topeka in 1872. Many carloads of building stone and flagstone were hauled to Kansas City by train. One of Horner's quarries was on 5th street, just west of the original Florence and where homes were later built. Another quarry was just east of Florence. Stone from both these quarries furnished much of the rock for railroad work along the Santa Fe.[16]

From Alexander Gardner's series, Across the Continent on the Union Pacific Railway, Eastern Division - 1867. Kansas State Historical Society

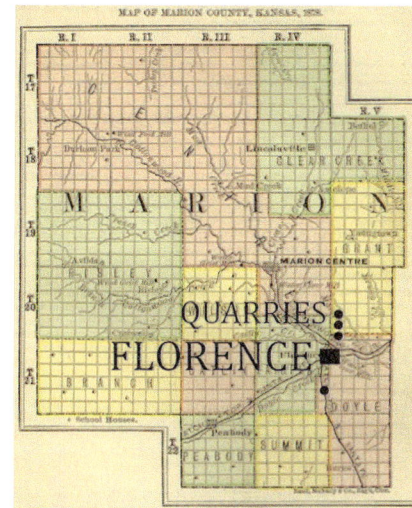

Map of Marion, County - 1878 - Quarry locations. KSGen Web Project on BlueSkyways

Stone from the Florence quarries had the reputation of being of the finest quality in Kansas in the 1880s.[17] Another quarry owner was Ed Jones who started a quarry northeast of town and another southwest of town. Railroad spurs were built to each of these quarries.[18] Morrison's Journal reported that there were five stone quarries in Florence in 1885.[19]

It is interesting that just as all the events that needed to happen for the threshing stone to make its way to being used in Kansas, the limestone quarry business was also just beginning and was ready to make the threshing stones as they were ordered.

Making: Who-How-Where

Who made threshing stones, how did they make them, where were they made? These are key questions to anyone with even a passing interest in threshing stone history. One holy grail of this project was to find a record of sale for the threshing stones, but so far this has not been found and most likely does not exist anymore.

To the question of who made the threshing stones, the answer is mostly skilled professional stone cutters who were employed by the quarries from which the stone was taken. However, some stories exist of the farmers cutting their own threshing stones.

The oldest record of making and cutting threshing stones, uncovered in this research, is in the following reference from the Russian encyclopedia *Historical-Geographic Dictionary of the Saratov District*, compiled by A. N. Mink, published in Saratov in 1898, translated using Google Translate from Russian.

"Up to 14 people were involved in making threshing stones, previously there had been more of them. In 1855-56, the Mennonites living on the meadow side, knowing where there is a suitable stone, ordered by model threshing stones. Two mountains, of which the largest Urakov mound, in the ravines and rivers found conchoidal stone, very strong, which goes to make the threshing stone. From that time began to be engaged constantly making them. They were originally made in 1 *arshin* length [2-1/3 feet] and 1 *arshin* of thickness and length was added after 2 *vershoks* [3-1/2"], the thickness of the same - with one side 12 *vershoks* [21"] in diameter, with another 11-1/2 *vershoks* [20 -1/8"], the narrow end of the stone when threshing is drawn to the center of the threshing circle. The stone price rose to 18 Rubles, but now for 3 and 4 rubles per stone. Taking them to the village they will sell for up to 10 Rubles per stone."[1]

100 Threshing Stones
Shipped to Kansas from the Ukraine on the
Red Star Shipping Line
at the Expense of the
AT&SF Railroad Company

Opposite Page: Cutting and shaping the stone is a work of art. Norman Epp facing a stone with a facing chisel.

The story of 100 threshing stones being shipped to Kansas shows up in many places. Origins go back to a book called *Wheat Country* by William B. Bracke, in 1950. "The Mennonites had things to learn too. With them they had brought a hundred so-called Russian threshing machines, going back almost to Biblical days, great stone surfaces, divided by deep troughs in the rock, a rod through the center with rotating shafts to separate the kernels from the chaff. Not a one of these horse-powered, devices was ever put to use in Kansas."[2]

The story seems to be a looping reference that has been re-quoted in many books, for example, *They Seek a Country* by David V. Wiebe, 1959. "At one time the Santa Fe company even chartered a Red Star ocean steamer to bring a shipload of Mennonite household goods and farm implements. These goods were brought to Philadelphia and then by rail to the settlers and all free of charge. The ship's cargo consisted of chests, crated wagons, plows, and Russian threshing stones."[3]

Here we see the Red Star Ocean Steamer added to the story. This story of stones being shipped here is however consistent with many families' oral history.

However most historians agree that these stones were not likely shipped from Russia but were all made here.

One piece of documentation speaks directly to the fact that the stones were made in Kansas which comes from

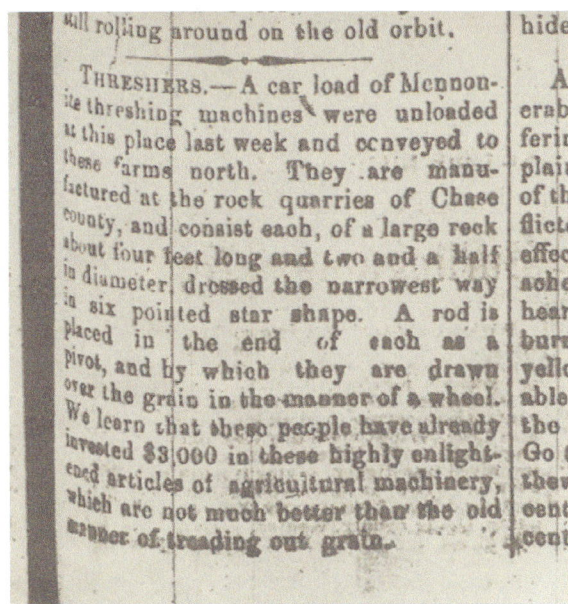

Microfilm copy of the Newton Kansan in 1874 on threshing stone delivery. MLA

The Newton Kansan, April 22, 1875. "A car load of Mennonite threshing machines were unloaded at this place last week and conveyed to these farms north. They are manufactured at the rock quarries of Chase County, and consist each, of a large rock about four feet long and two and a half in diameter, dressed the narrowest way in six pointed star shape. A rod is placed in the end of each as a pivot, and by which they are drawn over the grain in the manner of a wheel. We learn that these people have already invested $3000 in these highly enlightened articles of agricultural machinery, which are not much better than the old manner of treading out grain."[4]

This is the only reference I have found that specifically refers to the quarries of Chase County rather than Marion

Threshing Stone #16:

This stone sets on a concrete pad at the entrance to the Goessel Mennonite Heritage Museum parking lot. It is unique in that it has a 2 and one quarter inch square axle hole. This would imply that the axle was designed to turn with the stone and the pivot support would be in the frame. Irvin Goertzen said this stone is from the B. B. Wedel farm one quarter mile north of Alexanderwohl Church on the west side of the road.

Stone #16, Goessel Museum B

Threshing Stone #17

This stone is currently located in the Wheat Palace at the Goessel Mennonite Heritage

Museum. It is a stone that has a wrought iron frame. The ends attach to the axle with a simple loop. The front center flattens and widens to allow for a hole in the bar through which a ring is attached that can then

Stone #17, Goessel Museum C

be attached to the towing line. This stone shows considerable use, due to the enlarged axle hole and deep scrapes on each end. This may have come from past museum demonstrations more than from original use. It was donated to the museum by Roland and Marjorie Krause.

Goessel Wheat Centennial Celebration - 1974. Ben Boese

County in all other references. It should be noted however that the writer of this brief *Kansan* article has several facts that seem to be incorrect: "A large rock about four feet long and two and a half in diameter," as we now know most all are smaller, being 30" long and 2' in diameter, and "dressed the narrowest way in six pointed star," where all typically have 7 teeth.

The *Kansan* writer also offers his opinion on the threshing stone by speaking in what can be interpreted as a sarcastic voice, "These *people* have already invested $3,000 in these *highly enlightened* articles of agricultural *machinery*, which are not much better than the old manner of treading out the grain."

So is this true or is this false, how do we find out? I'll come back to that later.

Quarrying Stone

Quarrying the stone is the first step in making a threshing stone. As stated in other parts of the book, the limestone deposits in the Florence region of the Kansas Flint Hills are the ideal stone for building construction and equally suited for the unique threshing machine.

This limestone is ideal with the density of the stone making the appropriate weight, it is soft enough when first quarried to be easily shaped, and when dried is hard enough to last for many years.

The quarries are located where the layers of limestone are clear and accessible. It is likely that the stone quarried for the threshing stone was from a *bed* – a relatively uniform layer or ledge of rock which lies between two bedding planes or between different rock types - that is just slightly thicker than the diameter of the stone which is typically about 24 inches.[5] The horizontal strata give the stone maximum strength, so all threshing stones are made with the strata in a horizontal direction and parallel to the axle hole. This is evidenced by the fact that all stones that have been discovered with a crack running through them have been cracked in that direction.

I interviewed Norman Epp, a long-time sculptor who often works with Kansas limestone in his art, and is an expert in the forming and fabricating techniques used in the shaping of limestone. He believes the quarry could have culled the stone from a bed in a size very close to the final size that was needed to remove the minimal amount of stone to reveal the final product.[6]

In the 1870s there are several possible methods that would have been used to remove this chunk of stone. One method was to drill vertical holes, either by hand or possibly even at that time with steam-driven drills.

Then several methods could have been used to force these holes to create a crack that separated a piece of stone from the rest of the bed. Sometimes steel wedges would be used. These wedges and shims called *plug and feather*, would be pounded into the holes, ultimately creating a crack producing the appropriate chunk of stone.

Another method of the time, often used to make limestone fence posts, was to fill the holes with water and then allow freezing to crack out the chunk; however I think this was less likely. Dynamite could

Plug and Feather - Wedge and Shim tools used to break stone in quarries. Wikimedia Commons - Reiner Flassig.

have also been used, but this would have resulted in uncontrolled fractures, and is unlikely to have been used for masonry stones.

When limestone is first quarried it is in a much softer state and is fairly easily shaped. So, shortly after quarrying the stone, it would be given to a stone cutter for shaping. In the 1870s it was typical for the stone cutter to be employed by the quarry. These craftsmen spent a lifetime at their trade and were very skilled at cutting stone, evidenced by the beautiful limestone buildings and bridges found in central Kansas.

One possible stone mason of note from that time and working at the Horner Quarry in Florence was Oscar Johnson. His work is prominent throughout the area, including the little one room school house (National Register of Historic Places) just east of Florence, and he could have been one of the stone cutters who made threshing stones.[7]

Evidence also points to the fact that some Mennonite farmers may have made their own stones. Chuck Schmidt showed me a framed print of a threshing stone with a note on the back saying: "Threshing Stone - Shaped and carved with hand tools, the finished size is 2' diameter by 3' long, and a hole through the exact center for an axle parallel to the outer edges. The frame was not of 2" x 6" but split logs and horse drawn, as shown. All designed and and carved by Gerhard W. Nickel (Grandpa-Engineer), Johan

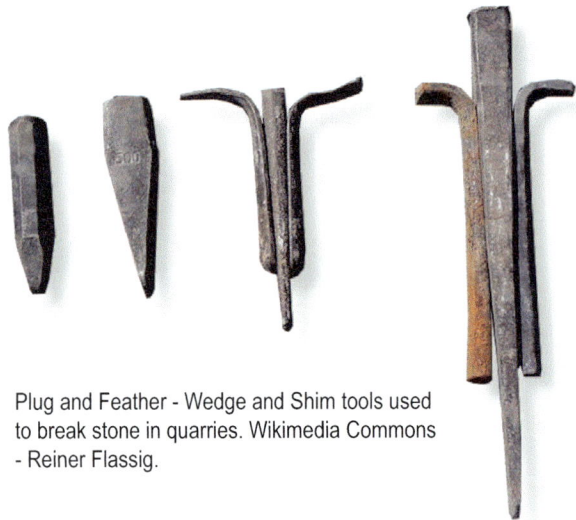

How to display a threshing stone?

I have seen many ways of displaying a threshing stone and many are still in excellent shape even if they have been outdoors for over 130 years. But it is my observation, that some will crack with continued exposure to freezing and thawing. It seems that it will crack along the horizontal strata of the stone, and that it may be less likely to crack if stored on its side rather than on end. It is also advisable to keep it out of the dirt by placing it on concrete blocks or a slab. Some individuals think the only proper way to display the stone is on its side like it would have been used.

Threshing Stone #18

I found this stone by surfing the internet, looking for local museums on-line to see if I could find any stones. I found 3 stones at the Hillsboro Mennonite Settlement Museum.

This stone has a wooden towing frame, made of wood planks that form a frame around the stone, with a cross piece in both the front and the back.

This frame is representative of the most common construction that can be seen in the oldest illustrations of threshing stone; however they were more likely to have be made of rough hewn wood or logs.

Stone #18, Hillsboro Museum

A sign hanging by the stones says, "The Mennonite Threshing Stone: The threshing stone with its seven ridges is a symbol of the sturdy Mennonite pioneers who introduced winter wheat to Kansas. This stone which is approximately twenty-three inches high and thirty inches wide, weighs about four hundred and fifty pounds.

When pulled over ripe wheat straw on a threshing

Scythe   Flail

Threshing Stone
(Ut Foa Steen)

THRESHING STONE
Shaped and carved with hand tools, the finished size is 2' diameter by 3' long, and a hole through the exact center for an axle parallel to the outer edges. The frame was not of 2"x 6" but split logs and horse drawn, as shown. A fanning mill was used to clean grain from all hulls and dirt.

All designed and hand carved

by:

Gerhard W. Nickel - Grandpa - Engineer
Johan Nickel   - Grand Uncle
Cornelius Funk   - Grandpa's in-law

(BRUDERTHAL)

Nickel (Grand Uncle), Cornelius Funk (Grandpa's in-law) - (Bruderthal)."[9]

With but a few exceptions, all the stones I have found have nearly identical dimensions and identical chisel marks that help qualify the construction of the stone. A few stones I have found show some variation from the norm. Stone #28, for example, is nearly perfectly fabricated far in excess of the typical stone. Another Stone, #34, shows unusually different strike marks on the faces of the stone indicating a variation from the normal mass-produced stones.

Round chisel marks typical on most threshing stones.

However, most all stones show very consistent fabrication techniques that can easily be observed on the faces of the stone. So what technique was used to shape the stone from a block of rock to a finished product that is the threshing stone?

Shaping Stone

In the interview with Norman it appears fairly logical to him as to how these shapes were formed and what tools were used that would yield the surface texture in evidence.

Knowing that the stone is fairly soft when first quarried and that tools will fairly easily shape the stone, it is most likely that the first step is to square up the ends of the stone. This process may possibly have been done with a saw, but more likely the stone could have quickly been dressed to parallel ends with the use of an *axe* or *tooth axe*, and then

Axe and Broad Bush Hammer.
Norman Epp

smoothed with a *toothed chisel*, and ultimately smoothed out with a *broad bush* hammer, a hammer with lots of small pyramid shapes on the face . The hammer is struck flat against the face of the stone and essentially pulverizes the surface so that small flakes and powder fall off. Repeated appropriate pounding ultimately reveals a smooth and flat surface when done by a talented craftsman.

According to Ray Wiebe, a wooden pattern called a *cull* was used to guide markings of the toothed pattern onto the end surface. Speculation is that the stone, while setting on end, was quickly first shaped into a cylindrical shape with the use of the axe and then again smoothed with the broad bush mallet to a relatively smooth finish. Once the stone is in the cylindrical shape, the "V" notches were then marked on the end of the stone and the notches would be roughed in with the use of the notched axe. Then using a round chisel the finishing of the V was accomplished.

Most all stones show a typical texture on the inside of the teeth. This texture is in the form of short repeated strokes against a round chisel moving the chisel forward slightly after each stroke. This method reveals a slight V-groove

Stone-shaping hand tools.
Marion County Historical Society

THE MENNONITE THRESHING STONE

The threshing stone with its seven ridges (the end view is shown in the logo) is a symbol of the sturdy Mennonite pioneers who introduced winter wheat to Kansas. This stone, which is approximately twenty-three inches high and thirty inches wide, weighs about four hundred and fifty pounds. When pulled over ripe wheat straw on a threshing floor by horses or oxen, it hammers the grains of wheat out of the heads. The stone was used by Mennonites in Kansas for a few years but was quickly replaced by more advanced farm machinery. About two hundred of these stones were made in quarries near Florence and Peabody, Kansas.

floor by horses or oxen, it hammers the grains of wheat out of the heads. The stone was used by Mennonites in Kansas for a few years but was quickly replaced by more advanced farm machinery. About two hundred of these stones were made in quarries near Florence and Peabody, Kansas." The provenance of this stone is unknown.

Threshing Stone #19

This stone now sets outside the barn at the Mennonite Settlement Museum in Hillsboro. When I first saw this stone I was very intrigued because it was a different color from any stone I had seen so far. It was a neutral gray with a few faint dark red streaks in the stone. I made a slight scratch on the surface, and paint flakes came off, revealing the tan limestone color underneath. This stone had been painted...I was curious.

Jim Juhnke recalled that two of these stones had

been painted red, white, and blue possibly for the U.S. Bicentennial. This was aslo confirmed by Ray Wiebe.

Stone #19, Hillsboro Museum

Obviously, this is one of those two painted stones; it has now been painted over with gray, and still has some of the red showing through. I believe the other painted stone to be Threshing Stone #60 now located in a monument on Main Street in Peabody, Kansas.

Threshing Stone #20

This is the third stone on the museum property; it sets on concrete near the sidewalk east of the house. This stone has the round tooth design. The provenance is unknown.

about ¼" deep and often about 1-1 ½" long in a generally straight line.

This groove was repeated over and over and is typically shown at a slight angle to the face revealing the direction that the carver worked the stone. These grooves often changed angles slightly as the cutter worked his way across the entire face. The fact that these grooves are not perfectly straight indicates that they were not made with a flat notch chisel but are typical of the rounded chisel.

It has been speculated that these small surface grooves were important to the threshing of the wheat, but I find no evidence that this had any effect. It was merely the result of the process of fabrication.

The wooden cull pattern was a negative shape, i.e. a hole in the board in the shape of the desired form. This cull could have been used to slide down the stone to check the accuracy of the cutting of the V notches. One of the "holy grails" of this project would have been to find one of these culls, but there is no evidence that any survived.

Finally, and I found this surprising, the very last step was drilling the

Cull or wooden pattern - Modeled and illustrated by author.

axle hole into the stone. The evidence that this was the last step exists in the dozen or so threshing stones that were never finished. They are complete except for an axle hole; they had become obsolete before they were ever finished.

The axle hole could have been put in using several methods. A drill bit existed that could have easily cut this hole. Or it could have been put in by using a star-shaped round bit repeatedly hit with a mallet and slightly turned on each impact to remove stone. The threshing stone is not a tool of great precision. The talent needed for the last step, centering the hole accurately and keeping it parallel to the sides of the stone must have taken great skill. It is most likely that the hole was drilled to the center from both ends.

Round Chisel and Toothed Chisel. Norman Epp

It should be noted that there are several variations to the axle hole. Most are about ¾" in diameter and go all the way through the stone. Some holes are larger, and this may be due to use. Some have a square hole. This implies that the axle rotated with the stone as opposed to the stone rotating around the axle. This last option is not a very durable solution, in that stone does not make a good bearing. A few stones have square recesses that do not go all the way through. The Kansas State Historical Society stone has a steel stub axle still wedged into this recess.

The consistency of the stones led to the deduction that most were mass-produced. Almost identical size, tooth geometry, and very consistent chisel marks, all are indications that these were mass-produced by the same group of stone cutters, making many stones at one time.

This conclusion is further endorsed by folklore that 50 at a time were ordered from the quarries. Also, David C. Wedel in the book *The Story of Alexanderwohl* says, "Threshing-stones had been made in Florence by a stone mason as

Axle hole can be easily bored with hand drill. KSHS permanent exhibit.

directed by Dietrich Gaeddert and Peter Balzer."[9] Unfortunately he does not cite a reference.

In interviews with Denver stone artist Norman Epp and Emporia stone artist Alan Tollakson, both agree that a stone could possibly be shaped by an experienced mason in 2-3 days of hard work.

Were some shipped to Kansas or were they all made here? Trying to find a definitive answer to this question was a time consuming but very interesting challenge. I took a 3-pronged approach to find out.

First I found that the Red Star Line was a shipping company that operated out of Belgium and is correctly credited for transporting Mennonite immigrants, and also many farm and household goods for the Kansas immigrants at the expense of the AT&SF. I was excited to find out that the Red Star Line is in the process of developing a new museum in Brussels. After several attempts, I finally contacted them through FaceBook, and asked if they had any shipping logs or bills of lading for goods shipped from the Ukraine to the US in 1874. The response however was, "Unfortunately, the passenger shipment and records in Antwerp from those years got lost."

Shipped Here or Made Here?

Stone #20, Hillsboro Museum

Also one of these 3 stones was at one time on the Tabor College campus, as reported to me by several sources. One of these stones was carved by Cornelius Duerksen, and used to be located in Wichita. Ray Wiebe recalls that it was moved from Wichita to the museum.

Threshing Stone #21

This stone is located by the Alexanderwohl Mennonite Church just north of Goessel, Kansas. It is most likely the most visible stone to be found, located on heavily traveled Highway K-15 next to a state historical marker "The Mennonites In Kansas." This roadside stop has been viewed by many people over the years. This stone is nicely displayed on a concrete pad with a round wooden frame attached, identical in design to the ones that were at Bethel College. This stone has a plaque located on the base that says, "This wheat threshing stone is the symbol of pioneer life and was used by the Elder Heinrich Banman."

Stone #21, Alexanderwohl, Goessel

Threshing Stone #22

This stone was first brought to my attention by David Linscheid, Director of Alumni Relations at Bethel College. We were talking about my research project when he said he had photographs of a tombstone in Elkhart, Indiana for Harry and Olga Martens made from an old threshing stone. These were

The Pennland, cargo ship of the Red Star Line Company - 1870. Heritage Ships of the Norway Heritage Collection

Further correspondence from them included, "During our research in Belgian emigration archives we have found some references to the Mennonite emigration. Mennonite emigrants are described as travelling in groups and being very wealthy, especially in comparison with other emigrants."

Additional maritime records research has found a ship by the name of *Pennland* built in 1870 that sailed regularly from Antwerp to New York carrying cargo, but it is uncertain if this is the ship that carried the Mennonite goods or just one very similar, so unfortunately, for now, I have

OFFICE OF

The Atchison, Topeka, and Santa Fe Railroad Company.

53 DEVONSHIRE STREET,

BOSTON, Jan. 1, 1875.

To the Stockholders of the Atchison, Topeka, and Santa Fé Railroad Company.

THE Directors herewith present their Report of the Company's affairs, together with the Reports of the Treasurer, the Superintendent, and the Land Department, for the year 1874.

The gross earnings for the year were . . . $1,250,805 69
The operating expenses for the year were . . 557,641 84

Net earnings $693,163 85
Operating expenses $44\frac{6}{10}$ per cent of gross earnings.

The year has been a prosperous one, as may be seen by a comparison with the returns of the previous year. In 1873, our gross earnings were $1,222,766.31, and our expenses $64\frac{2}{10}$ per cent, or $786,405.30. In 1874, our increase in net earnings was $256,802.84, or $58\frac{7}{10}$ per cent; and our expenses were reduced $19\frac{6}{10}$ per cent. The fact of such an increase in a year of general depression, we regard very important, as an index of what may be expected from the

AT&SF 1874 Annual Report to Stockholders. Kansas State Historical Society

reached a dead end.

Next I explored the records of the Atchison, Topeka and Santa Fe Company. Most of their records are located at the Kansas State Historical Society in Topeka. After spending many hours going through 1874-1876 boxes of records, receipts, reports and documents, including the 1874 "Annual Report to the Stock Holders of the AT&SF," no hard evidence could be found and surprisingly the annual report did not include any reference of shipping a significant amount of goods for the Mennonite settlers. I found many very interesting documents, but nothing to confirm that they indeed contracted with the Red Star Line-unfortunately another dead end!

One avenue left, "Geology." I thought that if I could identify the source of the limestone used, I might be able to tell if there were differences between stones, and maybe a geologist could tell where the stone came from.

Well the logistics of getting a stone to an expert or an expert to the stone is a challenging task.

I decided to ask permission of owners to allow me to take a chip off the old stones as geological samples. To my surprise everyone I asked was willing to let me take a hammer and chisel and get a chip. I tried to collect the most diverse samples I could. I obtained 11 chips, which is over 10% of known stones.

Author nervously chipping off a chunk of stone from the Bethel College threshing stone. Dave Linschied

I considered a theory that maybe the "Round Tooth" threshing stones were the ones that were brought over from the Ukraine and the "Square Toothed" were made here. I also noted that a few of the stones were a slightly different color, and also a few were a smaller size, so I decided to get chips representing all variations and from various communities.

Getting permission from individuals was one thing, but having the nerve to actually chip on these old stones was a bit intimidating. However, after having success on our own stone, my confidence grew...no stones were damaged in the process. I truly appreciate those individuals and organizations that gave me permission to collect these samples. So now what do I do?

Through word of mouth, and a bit of luck, I became acquainted with two highly experienced and respected geologists in the state of Kansas, who both helped me to get a better understanding of Kansas geology and to identify the origins of the threshing stone material.

I first met with Dr. Paul Johnston, Professor Emeritus of Emporia State University. After a detailed examination of each of the 11 stone chips as well as exemplary chips from the current Florence quarry, his observations were: "All chips show *echinodermata*, phylum of marine animals such as starfish, sand dollars etc. This is the shiny reflective surface in the stone. Two stones #5 and #34 show a different matrix, but the same fossils, this

actually relatives of mine. Harry would occasionally stay at our house when he visited Kansas on business trips for the Mennonite Biblical Seminary when I was a boy. Bethel College had received the photographs from Virgil Claassen who wrote "Attached are pictures of the threshing stone which is located in the Prairie Street Cemetery in Elkhart, Indiana. Harry and Olga Martens had this stone in their front yard of their home. Harry mentioned to me on a number of occasions that the stone would become their tombstone."

Stone #22, Tombstone, Indiana

Photo - Virgil Claassen

Additional information from Alton Longnecker, says that he had always understood that this stone had come from Buhler, Kansas. However later information from Harry's daughter says this stone has a long provenance.

Dee Miller said, "Harry and Olga are my parents. I have the background on the stone in mother's file. Her father brought it over from Russia. I do not have access to the file just now because we had a flood···in the files are pictures of the stone at Bethel. The stone came over from Russia with J. Reimer, it went to MCC in Lancaster, PA for 3+ years. It proceeded next to Elkhart, Indiana. It is in exceptional condition and is now at its final resting place as the grave marker for Harry E. Martens and Olga A. Reimer Martens." This stone was at Bethel college for a while and could be the missing stone.

Threshing Stone #23

This stone may never have been found had I not asked my mother Lydia (Schmidt) Ediger from the Buhler community if she remembered anyone having a threshing stone.

can be a variant within the stone even a couple of feet apart. Chip #2 has smaller fossils, but still similar. All the stones are the same texture, basic color, size of grain and similar fossils. This would all indicate that it comes from the Fort Riley Limestone strata (named for where this type of stone was first located) typical in Chase and Marion counties, a Rim-Rock that is often used as building stone."

He concluded with: "These chips might as well have all come from the same rock."[10]

Later I met with Will Gilliland, Adjunct Professor at Washburn State University and retired Environmental Scientist for the State of Kansas. His early observation was that these stone chips most likely came out of the Fort Riley strata, consistent with stone in Marion County. His conclusion was: "This stone is all Fort Riley Limestone, a Rim Rock bed also used on the Kansas State Capitol east wing. It has pockets of limonite which spread to give it the tannish color, but few fusilents, which would be more common in the Cottonwood strata. These chips are all from the same type of stone."[11]

Evidence is strong, based on my 3-pronged approach that all stones were made in Kansas and that it is unlikely that any stones came from the Ukraine, and if any did they no longer exist.

I appreciate the help of these gentlemen in providing me with a valuable education into Kansas geology, and in helping identify the probable source of the stone used in the threshing stones.

Chip #1
Goessel Community
Square Tooth

Chip #2 - BC Courtyard
Unknown Community
Round Tooth

Chip #4 - KM
Goessel Community
Square Tooth

Chip #5 - KM/Ratzlaff
Buhler Community
Square Tooth

Chip #6
Goessel Community
Square Tooth

Chip #10 - Buhler Monument
Presumed Buhler Community
Square Tooth

Chip #19 - Hillsboro Museum
Presumed Hillsboro
Round Tooth

Chip #20 - Hillsboro Musuem
Hillsboro Community
Round Tooth

Chip #34
Goessel Community
Square and Yellowish

Chip #66
Goessel Community
Square and Smaller

Chip #94
Hillsboro Community
Round Tooth

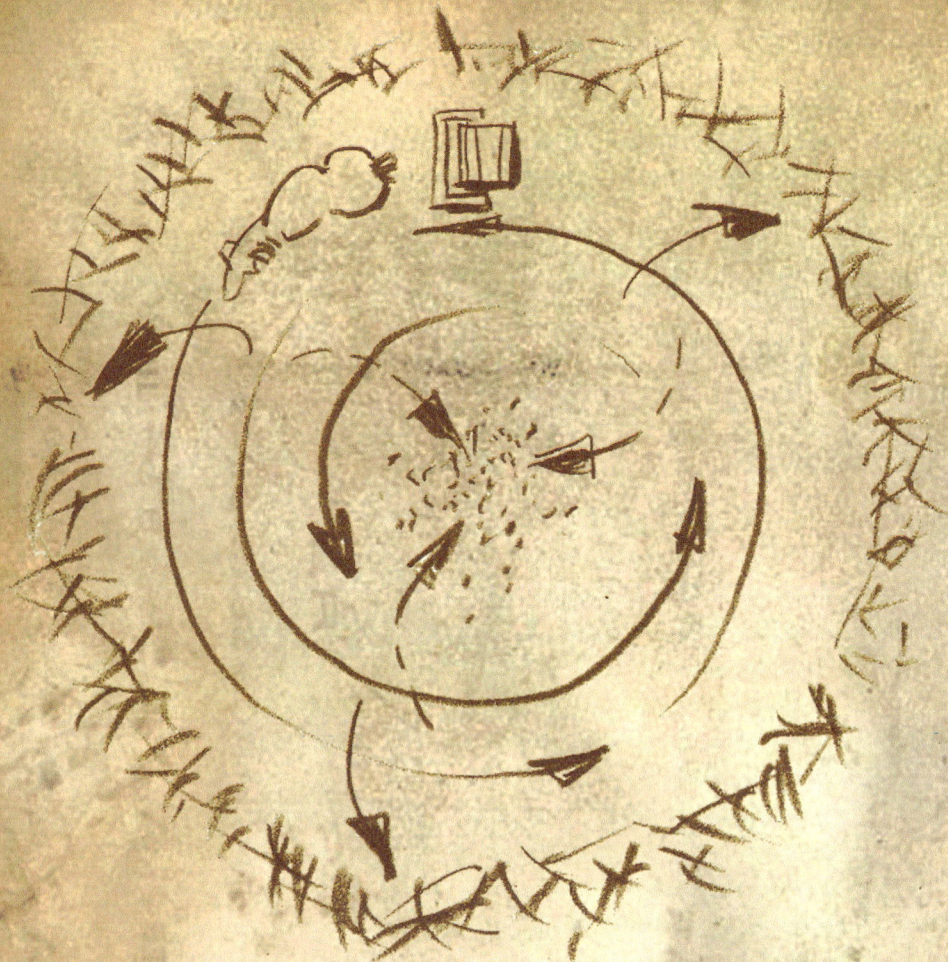
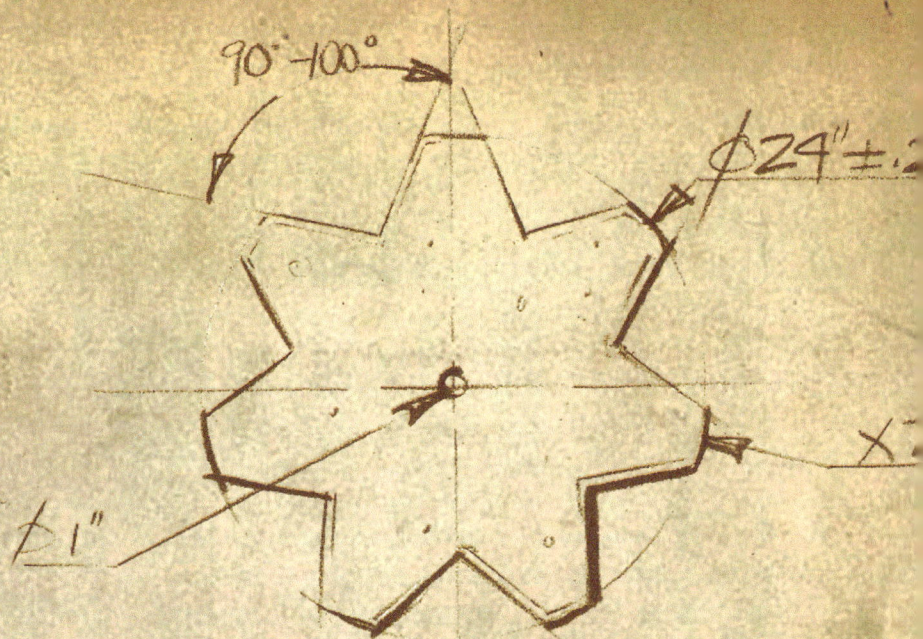

Design: Why Seven Ridges

One curious question that many have asked...“Why seven ridges?” This is a logical question given that none of us grew up with the use of the threshing stone, yet for many of us the shape is etched into our memory, even if we don't really know much about its design. What are the characteristics of the design that make it the way it is?

If you were to ask anyone to draw a threshing stone by memory, admittedly this is a small group of people in the world, they would probably sketch a circular shape with multiple notches around the perimeter that look something like the teeth on a gear. This is basically correct, but why is it this shape, this size, and why seven ridges?

There are numerous stone threshing machine designs that have been used since the Roman times. These design variations have evolved through the years, perhaps peaking and perfected with the widespread use of the threshing stone by the Mennonites of southern Russia.

I have found and studied hundreds of photographs of the threshing stones scattered about the steppes of the Ukraine, as well as about 100 stones that I have found in North America. These stones are remarkably similar in size and design, but have

Smooth Threshing stone in China - Photo 1976. Robert Kreider

fallen by the wayside over the last hundred years or so. Many were abandoned in fields, but fortunately many are in collections, museums, or on display in individual yards.

As stated in other parts of this book, the threshing stone has been used around the world at different times in history. There are design similarities to all, but why did it evolve to the stone that we find in North America?

Stone has been chosen as the medium most often used. However heavy wooden logs, wooden spiked devices, and in latter years in the Ukraine even concrete threshers have been used. Stone is used for many reasons; it's cheap, available around the world, is fairly easily shaped into a cylinder or cone, and has significant mass.

The weight of the stone is of great importance. If the thresher was too light, it would merely roll over the top of the straw, without having the mass to create

Opposite Page: Author's sketch - Showing how a threshing stone is designed and used.

the motion required to knock the grain from the head. Even though some stones found in China have no ribs at all, and others have only slight grooves cut into the face, the vast majority have significant "V" grooves cut into the side of the cylindrical stones.

Very few stones found have ever been weighed, and references to weights in writings vary greatly. In Inman, Kansas there are two stones on display by the Museum on Main Street. The sign by one says: "The seven-sided, cone-shaped wheels weighed from 360-720 pounds..."

I do not know if the author actually weighed any of the stones to come up with this weight range. Additionally, stones in Russia are claimed to be 100 kg, or 10-20 *pud*, about 450 pounds. I have collected the weights of three of the threshing stones from central Kansas. Our stone weighs 720 pounds. While moving John Thiesen's stone we ran it across the scale and found that it weighed 820 pounds. A third stone owned by a friend of mine was moved and weighed in the process at 760 pounds. Virtually all the stones found in central Kansas are the same size and shape so we must assume

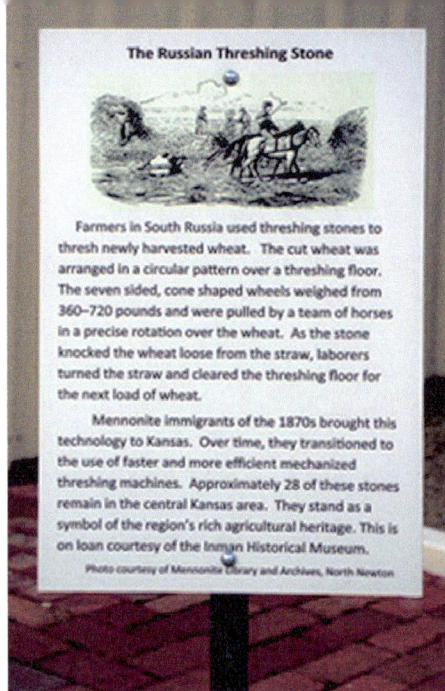

Sign on Inman Main Street - Says 360-720 pounds.

that they all weigh within the higher weight range of 700 to 850 pounds. So there is confusion as to why other persons' weights are considerably less than the ones weighed in Kansas.

It is obvious that various types of stone have been used around the world, evidenced in the color of the stones. All North American stones are made of local limestone, and the stones that remain in the Ukraine also appear to be mostly of limestone and later some made of concrete.

I have also measured many of the threshing stones that I have found, and I find it amazing that they are almost all the same dimensions to within about ½". The nominal size is 23-24 inch diameter and

Weighing threshing stone at the Harvey County Transfer Station.

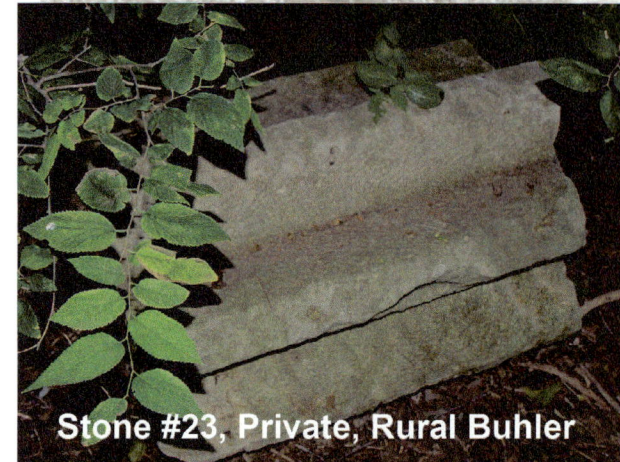

Stone #23, Private, Rural Buhler

She recalled that there used to be one on a Hoffnungsau member's farm on Dutch Avenue years ago. I contacted the son of this family who now lives on that farm, and he told me over the phone, "I think that there may be one in the trees behind the house." I stopped in one day and we did find it. I photographed it as it sat laying in the trees on its side, not visible to anyone who wouldn't know that it was there...a lucky find.

Threshing Stone #24

This stone came to my attention actually much later in my search but Stone #24 was reserved for a rumored stone in Greensburg that was never found, so I have inserted this stone here.

My friend Orlin Martens owns a heating and air-conditioning business and was doing some repair work in Buhler. He noticed a threshing stone displayed way back in middle of the large back yard. I made arrangements to photograph it, and I had a nice meeting with the elderly couple. They told me that this stone was on the farm

Stone #24, Private, Buhler

1 mile east and 1 mile north of Buhler next to the big lake originally owned by a Siemens family. It was there when they bought the farm. They later moved it to town when they retired. This stone is significant in that it is rare to find one tapered.

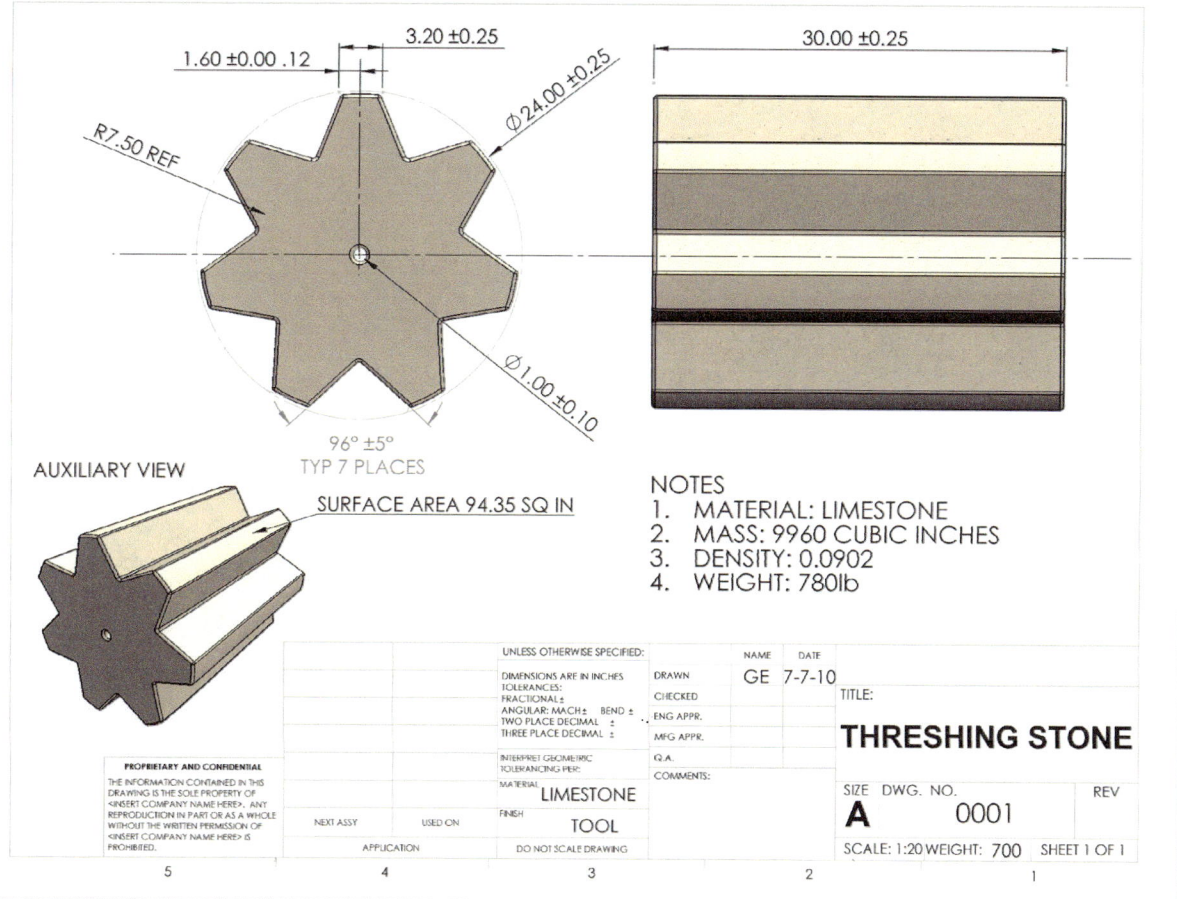

3.20 ±0.25

1.60 ±0.00 .12

30.00 ±0.25

R7.50 REF

Ø 24.00 ±0.25

Ø 1.00 ±0.10

96° ±5°
TYP 7 PLACES

AUXILIARY VIEW

SURFACE AREA 94.35 SQ IN

NOTES
1. MATERIAL: LIMESTONE
2. MASS: 9960 CUBIC INCHES
3. DENSITY: 0.0902
4. WEIGHT: 780lb

UNLESS OTHERWISE SPECIFIED:
DIMENSIONS ARE IN INCHES
TOLERANCES:
FRACTIONAL ±
ANGULAR: MACH ± BEND ±
TWO PLACE DECIMAL ±
THREE PLACE DECIMAL ±

INTERPRET GEOMETRIC TOLERANCING PER:

MATERIAL LIMESTONE

FINISH TOOL

PROPRIETARY AND CONFIDENTIAL
THE INFORMATION CONTAINED IN THIS DRAWING IS THE SOLE PROPERTY OF <INSERT COMPANY NAME HERE>. ANY REPRODUCTION IN PART OR AS A WHOLE WITHOUT THE WRITTEN PERMISSION OF <INSERT COMPANY NAME HERE> IS PROHIBITED.

NEXT ASSY   USED ON
APPLICATION   DO NOT SCALE DRAWING

DRAWN   GE   7-7-10
CHECKED
ENG APPR.
MFG APPR.
Q.A.
COMMENTS:

NAME   DATE

TITLE:

THRESHING STONE

SIZE DWG. NO.   REV
A   0001

SCALE: 1:20 WEIGHT: 700   SHEET 1 OF 1

Author's Engineering Drawing based on common dimensions of found stones.

29-30 inches long. However, examples from the Ukraine show more diversity of sizes judging from photos of these stones. I made a template of our stone and compared it to other stones, comparing tooth design, diameter, and looking for taper. I found remarkable similarity in geometry between all samples. Based on this collected data, I created an engineering drawing with tolerances. Most all stones found would comply, with noted exceptions for the round tooth design which is still based on the same geometry.

I have not had a chance to go to the Ukraine and measure stones there, but judging from numerous photos, the Russian stones are based on very similar geometry, with slightly deeper grooves.

It is obvious from looking at old illustrations of threshing stones from around the wold that many threshing stones have a taper. It's good physics for the stone to taper if you want it to travel in a circle.

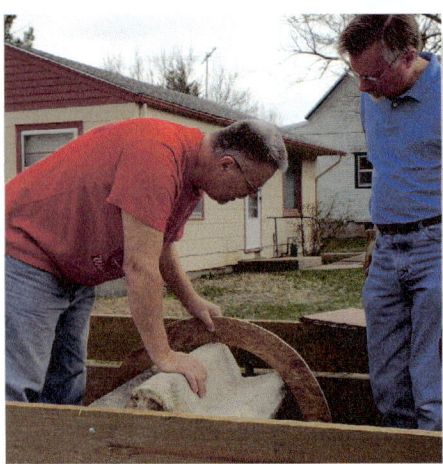
Glen and John check diameter and taper.

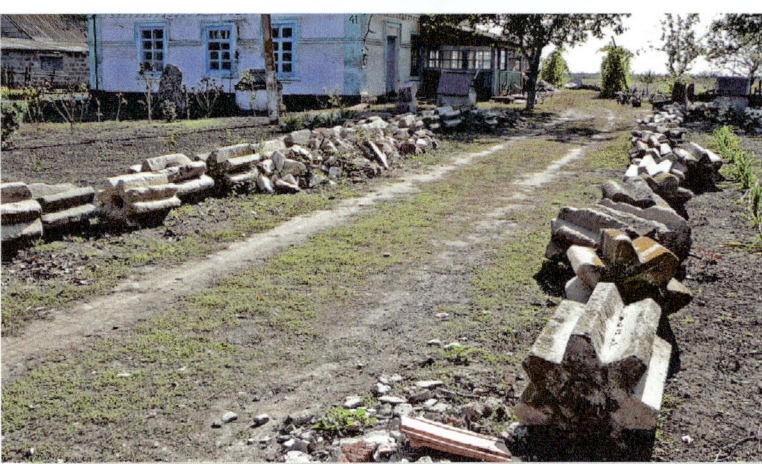
Russian stones show similarities to U.S. stones. Stan Hill

The first mention of this stone was from Rachel Pannabecker of Kauffman Museum. She said that Bob and Lorene Goering have a threshing stone. I called them and they no longer had it, as they had sold it to a neighbor. The stone comes from the Hillsboro area, on the property at the edge of town that her parents had purchased.

If you roll a cylinder it will go straight, but if you roll a cone (tapered) it will roll in a circle. Various degrees of taper will determine the circumference of the path of the cone. If you are interested in the math - to end up with a 50-foot diameter perimeter path, the outside diameter of the threshing stone needs to be 24 inches, creating a pretty obvious cone shape on a 30 inch long stone. If you want the stone to travel in an even smaller circle, a greater taper is required, similar to the ancient Chinese illustra-

tion showing a stone with extreme taper designed for a small threshing floor.

Many credible writings speak of the Mennonite threshing stone as having a conical shape; however my research has only found limited evidence of this. Only a few Kansas Mennonite threshing stone show obvious taper, and most stone images from the Ukraine do not show a taper.

"Taper"

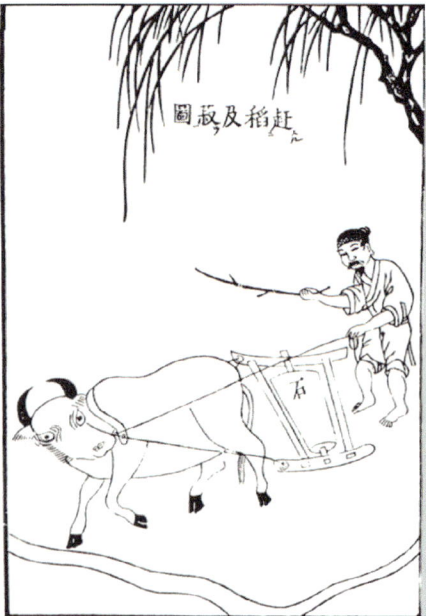
圖敁及稻赶

Chinese illustration of threshing stone shows extreme taper - 1637.

My theory is that when going around in a 30-50 foot circle on loose straw it really doesn't matter whether there is much taper, as there will be enough slippage on the straw that the horses will have no problem towing the stone in a circle. Additionally if the stone

Stone #25, Private, Rural Newton

They sold the threshing stone to an acquaintance who now lives out of state, but the stone is on a neighboring farm west of Newton. I finally contacted him and learned where the farm was and who the renter was now living on the farm. After many tries to contact the renter, I finally got to photograph the stone, with a cat cooling itself on the cool stone in the shade on a hot summer day.

Threshing Stone #26

I was at a funeral at Eden Mennonite Church sitting with individuals at the meal following the service, where I was explaining my threshing stone research project, and found that two individuals sitting one on either side of me both had threshing stones. Wow!

To my left was Karen (Unruh) Loucks, and she said they have the stone from her Unruh family in Goessel.

Months later I got an e-mail from Matthew Voth. "A few years back I helped move

Rare Kansas stone found with obvious taper both in diameter and in the tooth.

Stone #26, Private, N Newton

was to be moved from farm to farm it would travel down the road much better if it were not tapered. Again I think the Mennonite farmers, through trial and error experimentation just kept making what worked best.

"Why 7?"

The question of why the threshing stone has seven ridges has caused much speculation over the years and has generated several theories and numerous writings. By far the most common number

of ridges on a threshing stone is seven, even though historical illustrations show stones with no ridges, and others with as many as 60 or as few as 4.

There are many theories, but by far the most popular theory is that seven is a Biblical number. The number 7 has many numerological meanings in Christian and even more in Jewish tradition. Seven is considered the perfect number, symbolizing God's perfection, God's sovereignty and holiness. Genesis tells us of God creating the earth in seven days, the seven-day week is a reminder of our Creator, God blessed the seventh day and made it Holy. Both Old and New Testaments are filled with number seven symbolism.

"Holy?"

The late Wes Prieb of Hillsboro wrote that the seven ridges on the threshing stone are to remind us of the Holy Creator and serve as a reminder of seven Christian values which are essential to the development of a caring community.[1]

Another theory is that 7 ridges place a ridge opposite a groove, allowing the center of gravity to be lowered and therefore allowing for easier rotation. However, the physics of this proposition does not hold up, as the center of gravity is located about the center of the stone at the same location of the axle. I agree that the center of gravity will be at its maxi-

"Low CG?"

mum height when the stone is directly sitting up on a ridge, and lower when the stone is directly centered on a groove. This means the center of gravity will go up and down slightly and therefore the average CG is below the nominal radius of the stone. However, this lower average CG will provide no rotational advantage when the stone is being pulled. Actually this up and down oscillation motion requires greater energy input to continue the rotation.

This sub-motion, however, does provided a significant additional thumping action that is valuable in creating the knocking condition important in the threshing process. A similar principle is used today to help break up concrete for road repair.

"Odd?"

There is another possible reason why the grooves are in an odd number; this makes a ridge opposite a groove, thereby eliminating notches being directly opposite each other which would create a short path for a stress fracture to travel through the stone. Having an odd number means the shortest path from the bottom of two grooves will not pass through the axle hole and not follow the natural grain of the stone. Essentially an odd number is stronger and less susceptible to breakage.

"Syncopated?"

The next theory is also surprisingly accepted by many individuals. There is an odd number so that on traveling about the threshing floor, the tooth would hit in the gap left on the previous round. This theory falls apart for several reasons. In order for this to happen the stone would have to travel at a precisely predetermined radius, traveling at a perfect arc path. This of course is impossible to accomplish plus there is also uncontrollable slippage on the straw that would get the stone out of sync. Additionally if this theory were true it could also be done with an even number of ribs at a slightly different radial path. Therefore this theory is not viable.

"PSI?"

The next theory has to do with pressure. Some have speculated that the overall weight of the stone when supported by the surface area of one rib, calculated to be about 94 pounds per square inch, is the right amount of pressure to knock the grain from the head but not enough to crush the grain. This theory may, in part, be correct, evidenced in the fact that it does knock out the grain.

a stone to a house in North Newton from the old Unruh farm ground that I rent in the Goessel area." I went to Marvin Unruh's place on Ivy. I did not see any stone at this address, then I realized this was the stone now at Karen's house. This was a stone for which I had 5 separate leads that all ended up being this stone.

Threshing Stone #27

This lead came from the person sitting to my right at the funeral who said they had one on the farm where she grew up on Dutch Avenue. This is near where I grew up, and I used to hang out on that farm with fellow motorcycle riders, but I certainly did not remember the stone there.

Stone #27, Private, Rural Buhler

Photo courtesy of the owner

She said that it was still on the farm, and that I should contact her sister to make arrangements to photograph. Her sister sent me pictures of their stone as it sets on the yard now, but neither sister knows anything about the history of the stone.

Threshing Stone #28

This stone came to my attention one day while at Kauffman Museum. I asked Ilene Schmidt if she knew of any threshing stones. She said that she and Hartzel had a stone that they had given to Chuck Schmidt, their son. Chuck is an old friend of mine. He has it nicely displayed in a landscaped area. All that is known of the history is that it came from Hartzel's family.

What I found most interesting about this stone was the quality of stone cutting, it was by far the most precisely cut stone I have seen. There were no typical toothed chisel marks on the surface, but it looked more like it was cut with a smooth chisel. The corners were all very sharp and precise, unlike any other I have seen. I suspect

However, some have square teeth and some have round teeth so that theory does not entirely hold up either.

"Texture?"

Another design feature of the threshing stone is the surface texture. Most all stones show the remnants of chisel marks about 1/8 inch deep and about 3/8 inch apart. Some have speculated that this helps to thresh the grain from the hull, similar to modern combine thresher bars. This may or may not have a slight effect, but I believe it is nominal at best.

The "V" grooves that form the 7 teeth are cut into the stone with an internal angle of 90 to 100 degrees. These grooves are cut in an equally spaced radial pattern about the center line of the

stone, at approximately 51.43 degrees. With a nominal diameter of 24 inches, this equates to a circumferential dimension of about 75 inches. This appears to be the typical geometry used on most all stones. This geometry was easily reproducible in stone, appears to have worked well, yet was rugged enough to withstand years of use without breakage.

As a mechanical designer myself, I do not understand why the bottom of the V-grooves is made as sharp as it is typically cut. I see no reason that the bottom of the groove is important to the threshing process and the stone would be even more rugged with much larger internal radius.

Why do some have rounded teeth?

"Round?"

By far most examples of threshing stones have square teeth, but a few have rounded teeth. I have found 12 in North America, and I have not found any in the Ukraine.

Round tooth design.

The 1855 Petzholdt illustration (page 107) does show rounded teeth. Was this just an artistic interpretation or actuality? I am not sure why some threshing stones were made with round teeth, but can speculate that they were less susceptible to chipping along the edge.

"Square?"

Square tooth design.

Some have speculated that the teeth wore down to the rounded shape with use. But the radius is cut very precisely and many still show the original chisel marks in the radius, so this is not the reason.

One other characteristic that can be seen on many of the stones that remain in the Ukraine is a slight bump-out of about ½ inch thick and about 10 inch diameter on each end about the

center of the axle. I speculate this was included in the design to act as a built-in spacer to minimize the teeth of the stone rubbing on the towing frame when in use. It appears that all the stones that have this bump are concrete, not carved stone.

This detail has not been seen on any stones found in North America except for the stone found in North Dakota. The North Dakota #85 stone is also noteworthy in the fact that it is the only one found with only 5 teeth.

"Axle Hole?"

There appear to be several ways in which the axle hole was designed. The most common method in the U.S. is a 1 inch hole drilled through the entire stone. Either the stone would revolve around the axle or the axle would be inserted

Concrete Russian Stone with bump-out ends.
Courtesy Rachel Unruh Clark

Stone #28, Private, Newton

this one was not made in mass production like most of the others from the Marion County quarries.

Threshing Stone #29

Soon after I started this project, many people I talked to said I needed to talk to Jerry Toews. I soon talked to him and Leann and they have three stones setting under the trees in a semi-circle in their yard. He does not specifically remember where he bought some of these stones, but they were all purchased at auctions. One of his stones he had bought from Frank Pauls near Buhler. He also said he had a 4th stone at one time but is not sure who he sold it to.

This stone is the typical square tooth design and is in

Stone #29, Private, Goessel

excellent condition displayed with the others underneath a grouping of trees.

Threshing Stone #30

This second thresher has a heavy strap steel towing frame still attached to the stone. It is made into a "U" shape with the ends being wrapped around the axle, and a hole on the front

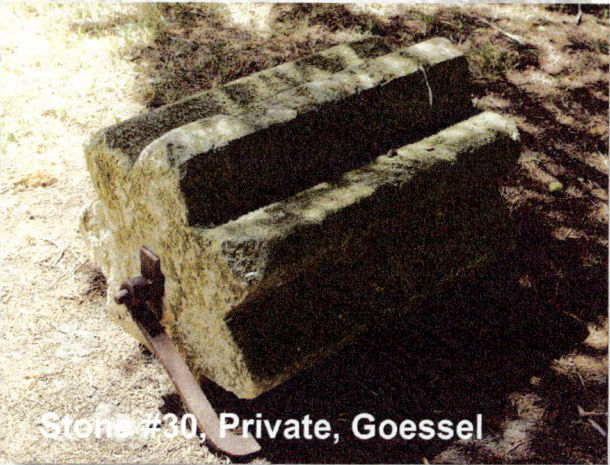

Stone #30, Private, Goessel

Square hole design - Moundridge residence.

tight enough to cause axle to pivot in the yoke. One stone in Goessel had a 1-1/2 inch square hole all the way through. And several stones had square pockets carved a few inches into the end, allowing for a short axle stub to be wedged into place.

Russian stone with steel straps near Neu-Balzer.
Photo courtesy Alexander Spack

The Russian concrete stones appear to have the axles cast in place.

Round hole design - Inman residence.

A few have a unique steel strap design that grabs around the stone.

If I have not completely overwhelmed you with details here is my conclusion on design.

Through simple trial and error over many decades, applying skills of observation, 7 ridges on a stone provided just the right amount of thumping action to accomplish the job, combined with the dimensions and weight of the stone, all worked together to thresh the straw with an appropriate force to knock the grain from the head, and with 7 teeth it rolls with relative ease. No taper on the stone is necessary because it tows in a circle just fine when on straw due to slippage and it will tow easily in a straight line.

The threshing stones are designed this way because it just works best.

"It Just Works Best?"

MENNO SIMONS.

Dit's Menno, die getrouw aan God en zyn gemoed,
Het Roomse bygeloof verschopte met den voet,
En dorst, hoe zeer gesmaad, ja zelfs in lyfsgevaaren,
Vrymoedig Christus Leer met mond en pen verklaaren.

te Amsterdam Uitgegeven by KORNELIS DE WIT, Boekverkooper 1743.

German - Russian and Mennonite History

Without a good understanding of Mennonite history one cannot explain how we ended up with a threshing stone in my front yard. The religious journey from Europe in the 1500s to the arrival of threshing stones in central Kansas in 1874 is a story of pursuing religious beliefs with multiple migrations to keep the practice of those beliefs alive.

The history of the Mennonites is a history of a people "on the move – for conscience sake." They fled from one city or state to find refuge in another. Their story is one of wandering...stopping here and there, sometimes for a few hundred years and sometimes scarcely for a generation.[1]

"For more than 400 years, the majority of the Mennonites have been people of the soil. Their almost fanatical devotion to farming can be traced back to the Reformation, when the sect was born in Holland and Switzerland as a product of the Anabaptist movement."[2]

All through the Middle Ages the desire to reform the church was strong,

Martin Luther posting his 95 theses in 1517 - Painting 1872 by Ferdinand Pauwels. Wikimedia Commons

and many proposals for reform were heard. This movement evolved into the *Reformation*, a backward-looking movement to return to the virtues the church had once possessed.[3]

October 31, 1517 is given as the date of the beginning of the Reformation with the posting of the ninety-five theses at Wittenberg Church by Martin Luther. This movement, however, was not enough for some individuals who, as a group, became known as *Anabaptist* (baptized-again), promoting "believers' baptism," a separation of Church and State, and the doctrine that religion is an individual heart-driven experience.

Opposite Page: Menno Simons print - J. C Philip - By Kornelis De Wit, 1743 - Copyrighted 1887 Welty & Sprunger. Author's collection

Menno Simons was the outstanding Anabaptist leader of the Low Countries during the 16th century, where his followers became known as *Mennonites*.[4] He was devoted to the beliefs of the earlier Anabaptists with a strong doctrine of nonresistance, which turned his followers into easy targets for persecution. Becoming victims of martyrdom, many were killed; they turned the other cheek and did not resist their enemy.

So they sought a place of refuge, where they could live in peace and continue to work with their farming skills.

In response to the invitation of the Polish and Prussian nobility, those that were not willing to give up their faith decided to preserve it by emigration. They settled by the thousands in the delta of the Vistula River in the early-to-mid 16th century. "They seemed to work a special magic with plow and harrow; with their knowledge of dike-building and farming marshy land, they transformed the delta from a swamp to a productive region of rich farms."[5]

Mennonites from across Europe were a part of the migration to Prussia, including the Low German and Swiss Mennonites.

For a period of more than two hundred years, 1540 to 1789, the Mennonites enjoyed the protection of benevolent rulers in their territory. But in the end, harassed by their neighbors and fearing the growing militarization of the Prussian state, the Mennonites began to search for another nation which would guarantee them the right to follow their consciences and practice their faith.[6] They were also restricted in the acreage of land available to them.

So again they looked for a place to farm with greater economic opportunities and to practice their religious beliefs in peace. It's hard to know which was the most significant driving factor.

The story of the next Mennonite migration begins with the invitation of Catherine the Great of Russia, inviting Germans and other Europeans to settle and occupy the lands vacated by the Turks in southern Russia.[7]

Realizing that agriculture was the backbone of national prosperity, she became interested in settling her newly

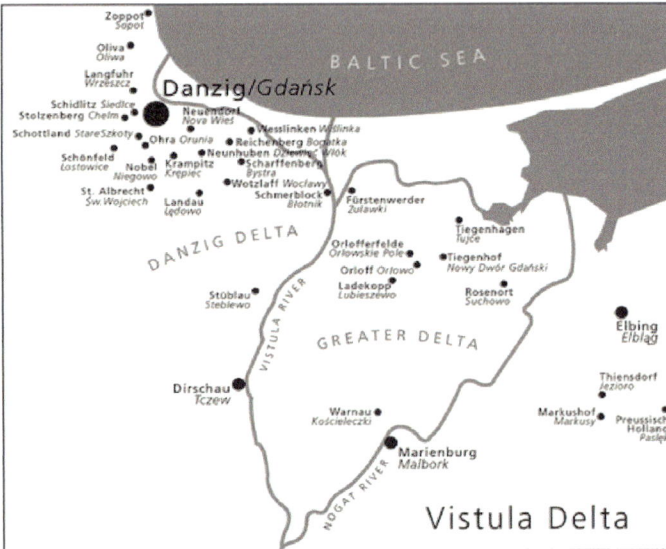

Mennonite settlements in the Vistula Delta and River Valley - Now part of northern Poland. MLA

center for a clevis, very simple and very strong.

Threshing Stone #31

This third stone is noted for not having any axle hole, so it could have never been used.

Stone #31, Private, Goessel

Threshing Stone #32

This stone came to my attention from a conversation that my wife was having at our insurance agents' office, where Elaine Harms overheard Karen talking about my project. She said that her husband Jim had a threshing stone, and also that he had done some research on threshing stones many years ago. I contacted him and he said he did have a stone but it is buried in his warehouse. He later sent me a photograph. He had bought the stone from Glen Pestinger, an avid antique collector in Hillsboro.

Stone #32, Private, Newton

Threshing Stone #33

This lead also came to me through a conversation my wife was having with Valerie (Loganbill) Klaassen. She said her par-

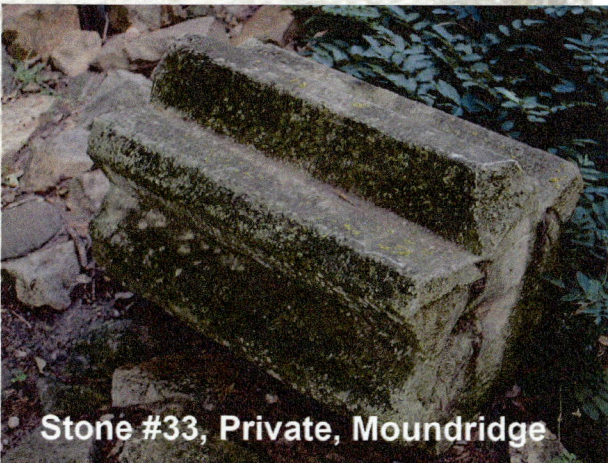

Stone #33, Private, Moundridge

acquired lands with efficient and thrifty farmers. Wherever people were hampered in their religious liberties, or were dissatisfied with economic or political status, attractive offerings were made.

Thus, progressive, agricultural colonists were given free land, free transportation and support for some time, exemption from military service and other civil obligations, religious freedom, and the autonomy to establish educational and local political institutions according to their desire.[8]

The first groups to leave Prussia, sold their land to others who were then able to stay with more land available. Along the coast of the Black Sea and the Sea of Azov, settlements of German farmers appeared. They were Reformed and Lutheran Pietists, Catholics, and Mennonites.[9]

Catherine II in 1783 granted special privileges to a group of German Mennonites if they would come to Russia and colonize what is now the Ukraine. The Mennonites were known as a very thrifty and industrious class of people, among Prussia's better class of agriculturists. Catherine hoped that the Mennonites would serve as a pattern for the indolent tribes scattered over Southern Russia and that they would intermarry with these tribes, but she was destined to be disappointed, for the immigrants kept to

Low German and Swiss Mennonite settlements in the Ukraine. Wikimedia Commons - United Nations Cartographic Section

themselves and seldom intermarried with the Russians.[10]

The Old Colony was settled by 228 families in 1789 at the confluence of the Chortitza River and the Dnieper River.

The second colony, located some 100 miles to the southeast along the Molotschna River, was founded in 1803. The founders of the Molotschna Colony were mainly experienced farmers who brought more implements, furniture, money and other possessions with them. These, therefore, became the most successful settlers of the Ukraine.[11]

The migration from Prussia to Russia spanned decades. During this journey one group met Czar Alexander I of Russia, who wished them well (German: "wohl"), prompting the naming of the new village, Alexanderwohl.[12]

The migration to Russia brought together Mennonites from different localities in Prussia, producing considerable variation in the Mennonite mosaic.

Here they engaged in agriculture, raising mostly wheat, as the grain best adapted to that locality. At first they raised the *soft* spring wheat. It was not until about 1860 or thereabouts that the *hard* winter wheat which was raised exclusively in the Crimea, was introduced into the Milk [Molotschna] River colonies.[13]

The Mennonites were very successful at turning the steppes into fertile and productive farmland. While most of the Mennonite population were subsistence farmers, they also excelled in agricultural businesses, concentrating on the processing of wheat and the exporting of wheat to European markets, as well as the manufacturing of agricultural machinery such as mowers and plows.[14] Many Mennonites accumulated much wealth.

The total Mennonite Russian-German population of "New Russia" was about 40,000 by 1869, half of whom lived in the Molotschna Colony, while the number in all of the Russian Empire was probably not over 75,000.[15]

Farming was, to start with, done on a small scale. They raised rye, oats, barley, wheat, potatoes, and vegetables.

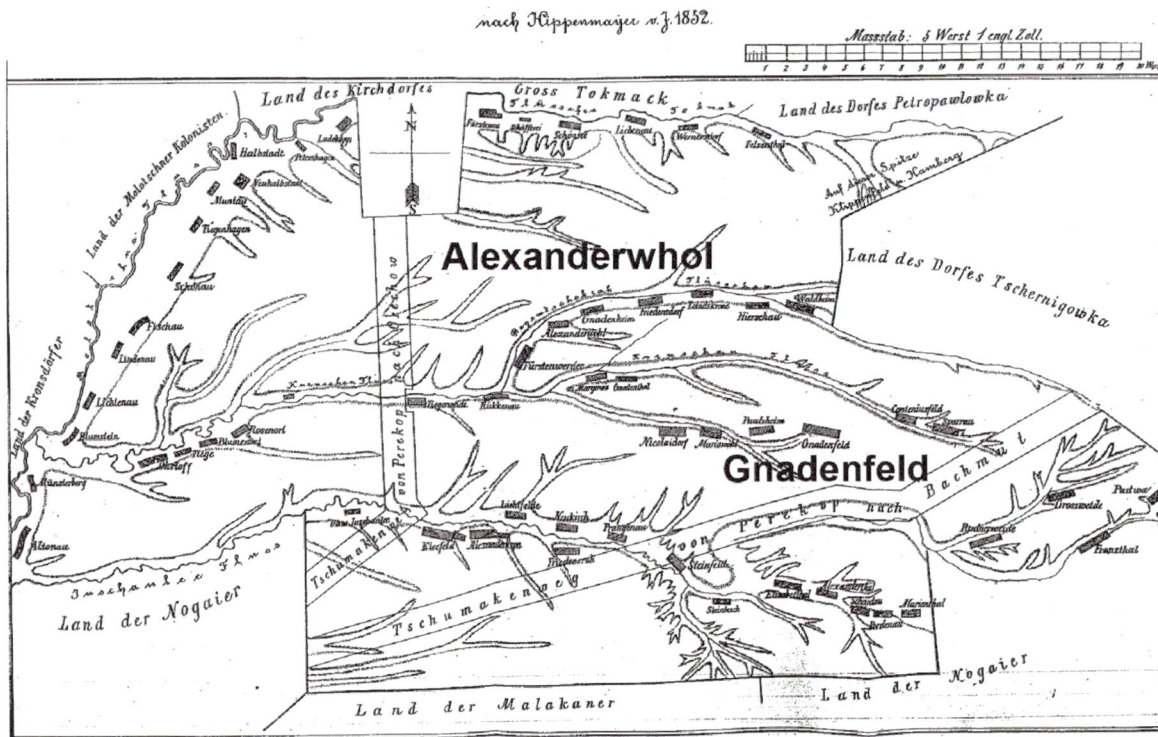

1852 Molotchna Colony Villages. Wikimedia Public Domain

ents had a stone at their house in Moundridge. I contacted Varden Loganbill and made arrangement to photograph the stone. He had purchased it at a farm auction many years earlier. The stone is now kept by Blair Loganbill.

Threshing Stone #34

Stone #34, Private, Rural Newton

This stone was discovered when Karen was attending a birthday party for some of the women from our church that get together for July birthdays. Last year they met at Susan (Schmidt) Rhoades' house and she told Karen they had one from her Schmidt family. I immediately made arrangements to photograph it.

They use it as a base for a bird bath, beautifully displayed in their back yard. The stone is a bit yellower than most stones and the chisel marks are much finer than typical. This makes this stone a bit unusual and unknown as to its origin; it also has a slightly different chiseled texture. The stone comes from her father, Dr. Herb Schmidt's farm northwest of Goessel.

Threshing Stone #35

This stone came to my attention at the Goessel Country Threshing Days; Milton Goertzen said that a stone may still be at Linda Bartel's home place, now owned by Rod Peters. I made arrangements to photograph this stone at the farm; it was nicely displayed on a concrete pad just east of

Stone #35, Private, Rural Goessel

Beautiful Russian Mennonite farm in Tiege, Molotschna, Russia - circa 1900. MLA 2003-0103

Drought was a severe handicap and made farming risky. By means of summer fallow (planting a crop every other year), productivity of the soil could be increased and drought was overcome. A foreign observer who noticed the difference between a Mennonite-owned field that had been summer-fallowed and one that had not was heard to have exclaimed: "Why? Do the Mennonites have a different God?"[16]

"The rapid development of the Molotschna Colony is due also to other factors. It cannot be imagined without the *Agricultural Association*, the far reaching influence of which was due to a man named Johann Cornies, where his estate became a model in all branches of agriculture and an experiment station."[17]

But eventually big estate farming was an outgrowth of the wheat revolution. One-third of all of the land owned by the Mennonites belonged to three hundred eighty-four families, the largest estate consisting of fifty-four thousand acres. But three-fourths of the Mennonites farmed about 200 hundred acres each and lived in the traditional villages of thirty to fifty homesteads. They are the pioneers that made the Ukraine the "Granary of Russia."[18]

But the success and continued prosperity of the Mennonites had also caused a feeling of jealousy among the native Russians and the tribes in southern Russia and they wanted the many special privileges granted the Mennonites to be withdrawn. The changes were all driven by the Russian elite's recognition that they had to modernise their country, get rid of serfdom, reduce special privileges and get people closer to being of equal legal status.

Simplified migration map of many Swiss and Low German Mennonites. Wikimedia Commons - Europe Map 1919

The treatment accorded the Mennonites by the Russian Government, up to 1871, was all that could be desired. The agreements made in the days of the Empress Catherine were faithfully kept.

Everything went well until the government announced its intention of enforcing universal military conscription. Against this the Mennonites protested and ten years was given them to yield or leave.[19]

The frontiers of the Ukraine were now settled and established, and the nationalists in the government were dissatisfied with the slow process of the "*Russianization*" of the foreign settlers. Assimilation was to be sped up by restricting the semi-independent status the colonies had enjoyed, especially in matters of school and public administration. [20]

This forced many to consider their options. The loss of exemption from government conscription, considerable population growth within the Mennonite community, and especially the lack of additional land options spawned the desire for another migration to find land and religious opportunity.

So yet again they looked for a place to farm and practice their religious beliefs in peace.

You may read further details in Article 8 about the railroads and in Article 9 about picking Kansas, where we see that

the house. He said as far as he knows this stone has always been on this farm.

Threshing Stone #36

I was told about this stone at Goessel Threshing Days. The stone is at the LaVern Goossen farm. It came from the Herman Voth family. I also heard that same day from Verny Voth that his grandpa had a stone; this is also that stone.

The stone was neatly displayed in the farm yard on a concrete pad, it has a partial crack and no axle hole.

Stone #36, Private, Rural Goessel

Threshing Stone #37

When I was at Rod Peters' farm he told me of another stone on a farm a few miles from his place, so I went there next and took a picture of the stone in the center of the yard before I had permission. Later I contacted the owner, who said the stone had always been on the yard while he was growing up. The stone is in rough condition, setting on top of a concrete block.

Stone #37, Private, Rural Goessel

Threshing Stone #38

I heard at the Goessel Threshing Days that Virgil Unruh had a threshing stone. I went to their farm near Hoffnungsau Church and learned that they actually have two stones. One threshing stone is setting by the driveway in a nicely landscaped area. It was from his family,

Typical Mennonite Farmstead in Molotschna with combined house and barn structure - Note threshing stone in the middle of the picture - circa 1900. MLA 2003-0228

many chose the opportunities in North America that they believed provided the best options for the next generations.

Once again the Mennonite migration was set to begin. After considerable exploration, and tempting offers from the railroads of the Great Plains, many started making plans for the great migration to the United States and Canada.

The threshing stone story is not only a story of the Mennonite migration, but also includes the Volga Germans who also ended up in Kansas. However the massive Mennonite migration as a collective body is a unique one and ultimately they were the main group that ended up using the threshing stones in Kansas.

It has always impressed me that as the Mennonites emigrated they left much behind, but what they also always took with them was their commitment to their beliefs, their farming skills, their business prowess, and even their language.

It's amazing that some Swiss and certainly the Low German languages have continued to survive all these hundreds of years from country to country, as the Low Germans hung on to their traditions. *Plautdietsch* is the distinct Mennonite Low German language which developed over a period of 300 years. Many Plautdietsch words show up in this book.

Unfortunately the language is finally coming to an end for most people. My parents' generation still spoke primarily Plautdietsch on a daily basis, but my generation has not learned to speak it.

However, I truly think many of the principal characteristics the of Mennonite families are still very recognizable in Mennonite families today, wherever they live now.

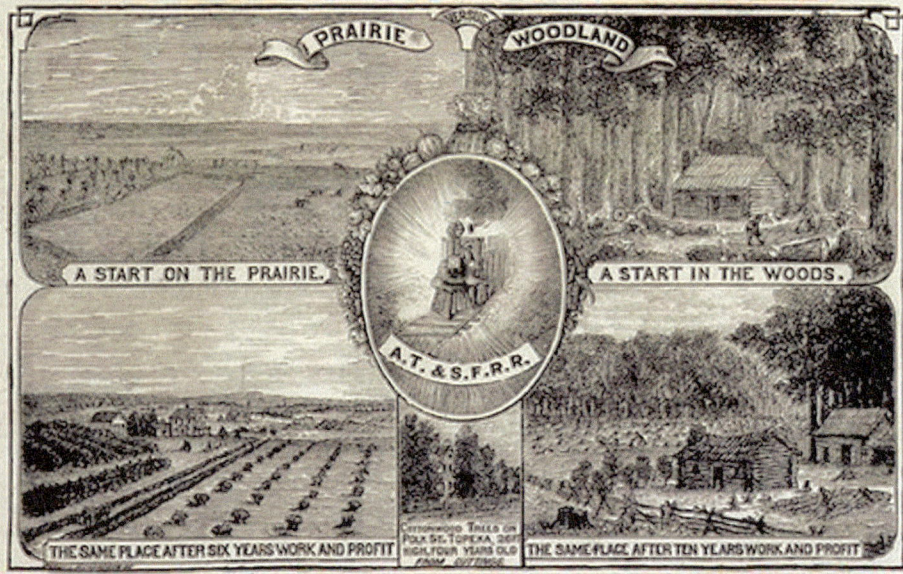

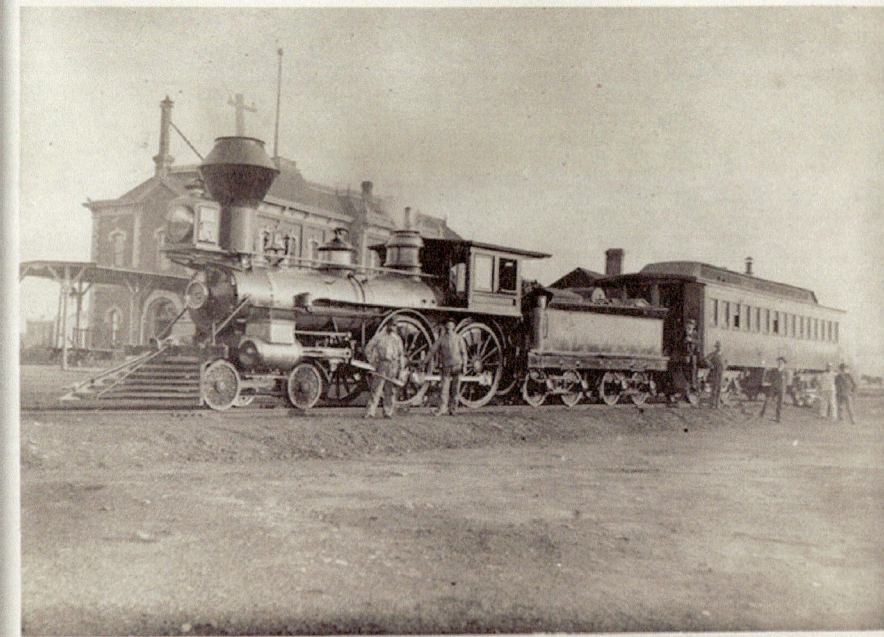

The Railroad Influence

If not for the railroad, threshing stones would not be found on the plains. Consider the times and it all becomes clear how the railroads played a huge part in the development of the Great Plains. In the mid-19th century the prairies were ripe for economic development, and the railroad needed people and commerce to be a viable venture. They made attractive offers to farmers to fulfill this need.

The U.S. government wanted the lands to be developed into productive real estate with loyal citizens to support the newly established direction of the country. It is impressive how the foresight of the railroad tycoons inspired ways to capitalize on this unique opportunity to shape the country and to profit in the process.

The times were right, the Civil War was over, Kansas was ready for its new identity. Many new Kansas citizens were actually pushed into the state to swing the tides of history. The U.S. government had all but cleared the state of the native people by forcing them off the land, first, with the elimination of their primary resource, the buffalo. Buffalo were essential to the lives of the Native Americans; their entire survival structure was based on the food and resources derived from the buffalo for daily living and trade. The elimination of the buffalo made life impossible for them to survive and gave leverage to the government to relocate them to territories to the south.

Additionally, the buffalo were a

Opposite page: 1870s Atchison, Topeka & Santa Fe poster used to lure settlers to the Kansas prairies. Note: Good Soil for Wheat, Corn and Fruit. Kansas State Historical Society

Opposite page: 1874 AT&SF Logo.
MLA - Dietrich Gaeddert archives

Opposite page: Circa 1870s photo of engine No. 5 of the Atchison, Topeka, and Santa Fe Railroad.
Kansas State Historical Society

The Far West. Shooting Buffalo on the Line of the Kansas-Pacific Railroad, "Frank Leslie's Newspaper", V. 32, No. 818 (June 3, 1871), Page 193. Library of Congress - Prints and Photographs Division

nuisance to the railroad companies that wanted to eliminate them from the path of the rail. Many men including Bill Cody, known eventually as Buffalo Bill, helped rid the state of this nuisance and paved the way for the railroad.

The world infrastructure had been based on moving goods via waterways. The highways of the times were rivers and lakes. The states to the east were carved into their shape mostly by the geographical influences of mountains and water passages.

But due to changes in technology, with the development of steam power, the railways could go anywhere and be free of the constraints of natural water routes. "Of course as the resources of the country are developed, lines will be made in all directions, but without doubt those that run east and west will be the ones first undertaken, and are the only ones that will command attention in the east." stated a newspaper in the 1860s.[1]

The U.S. government also influenced the push to build the rail system with several major acts. "To link these widespread regions with one another and with eastern markets, fast and reliable transportation was needed. The railroad was the ready and obvious answer. Kansas businessmen and political leaders even before the Civil War dreamed of rail systems which would connect their infant cities with every place of

importance in the nation. However, they soon learned that private enterprise alone could not finance such costly undertakings. Particularly in those areas where settlement was sparse and investment capital was slow in yielding returns, it was found that governmental assistance was necessary. This came in the form of land grants, and sometimes cash, from the federal and state governments, and from county, city, and township bond issues, which were exchanged for railroad stock and a promise that the company would build their way."[2]

"The said company is hereby authorized and empowered to survey, locate, construct, complete, alter, maintain, and operate a railroad, with one or more tracks, from or near Atchison, on the Missouri River, in Kansas Territory, to the town of Topeka, in Kansas Territory, and to such a point on the southern or western boundary of said Territory, in the direction of Santa Fe, in the Territory of New Mexico, as may be convenient and suitable for the construction of such railroad; and, also to construct a branch of said railroad to any points on the southern boundary of said Territory of Kansas in the direction of the Gulf of Mexico."[3]

To create the need for railroads many things had to be set in place. There was a need for people to create commerce, a need for farmers to grow crops, the need for towns to have merchants to

Stone #38, Private, Rural Inman

originally owned by David G. Unruh.

Threshing Stone #39

The second stone was located in the backyard and came from Virgil's uncle's farm owned by Marie and Albert Enns; it was on the farm when he acquired that property.

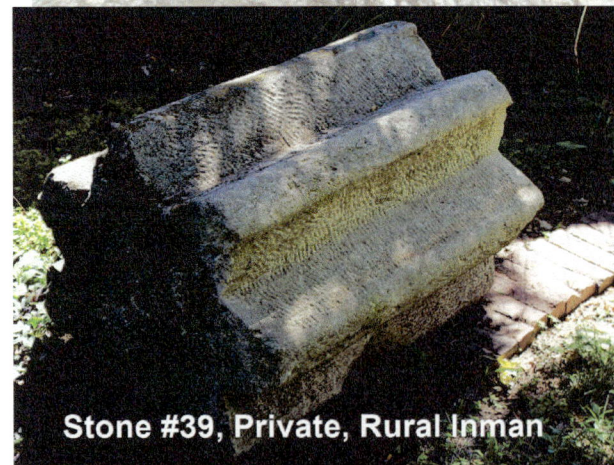

Stone #39, Private, Rural Inman

Threshing Stone #40

This was the other stone that Fern Graber had at one time in North Newton. She wanted it to go back to the farm that it came from, and this is where it is now located. The stone is displayed next to the sidewalk in front of the house.

Stone #40, Private, Rural Moundridge

Threshing Stone #41

This stone was mostly a lucky find. It reminds me that for all the stones that I have found there must be many that I will never find.

While at the Goessel Country Threshing Days, Arlin Buller told me that when he was teaching in Salida, Colorado,

sell goods, the need to supply the merchants with the goods, and also the need to take the produce of the lands to the populace that still lived back east. The need was created and the solutions followed.

"Acts of congress set aside eight and a half million acres of Kansas prairie to promoters on condition that railroads be built through the territory. The successful accomplishment of this task by 1872 gave the two giants, the Kansas Pacific and the Atchison, Topeka, and Santa Fe the right to claim seven million acres in alternate sections 20 miles on both sides of their rights-of-way. The railroads had not only acquired good agricultural land, but had also contributed, in no small degree, to the creation of a more civilized urban environment."[4]

The winner of the largest rights across central Kansas was the Santa Fe Railroad. This company was founded by Cyrus K. Holliday. His great win over the other railroads provided the AT&SF with access to some of the best farmland in the country.

Cyrus K. Holliday, a founder of Topeka, Kansas, and the father of the Atchison, Topeka and Santa Fe Railroad. Wikimedia Commons

A correspondent of the Topeka *Commonwealth* in 1869 said, "The variety of scenery which is found in Kansas far surpasses almost any country we have ever seen. . .we often think that there must have been some mistake as to the location of the garden of Eden."[5]

The right to develop the line came with great expectations and deadlines. "Holliday first conceived the idea of building a railroad in 1859 for the following reasons: to replace the slow, dangerous and overcrowded Santa Fe Trail; to terminate the Chisholm Trail at a point south of Abilene; to connect with the Missouri, Kansas & Texas Railroad at Emporia; and to handle the coal business from the coal mines at Burlingame."[6]

Mr. Holliday also drafted the Land Grant Bill which was passed and signed by President Abraham Lincoln on March 3, 1863. The Land Grant Bill ordered the State of Kansas to permit a railroad to be built from Atchison by way of Topeka, west to the state line in the direction of Fort Union and Santa Fe, New Mexico Territory, to reach the state line not later than March 3, 1873. This bill also stipulated that Kansas would give the railroad every alternate section of land designated by odd numbers on each side of the right-of-way which could be sold to help build the railroad.[7]

Eventually the line was started. There is some question as to when the first spadeful of earth was turned marking the beginning of construction on the

Santa Fe. Railroad, tradition and several historians say the date was October 30, 1868.[8]

The first 27 miles of track was laid in 1869 running from Topeka to Burlingame. Construction paused until new investors from Boston took control and in 1870 the line was extended to Emporia.[9] Then Santa Fe track was laid through Florence and Peabody during the spring of 1871 and continued on to Newton.[10]

The laying down of the roadbed for the rail lines was a major influence in opening the quarries in Marion County and created tremendous new business opportunities.

Abilene had boomed since 1867, with the thousands of cattle that had been driven there every year up the Chisholm Trail to then be shipped off to the East. But by 1871 the run was ended with connections closer to the south--first to Newton and then Wichita.

"In 1872, the Atchison, Topeka, and Santa Fe had finished its tracks through Kansas to the Colorado line, thereby earning a grant of three million acres. A map of this area in central and western Kansas resembled a checkerboard, railroad land and government land being located on alternate sections."[11]

"The change that occurred in the unruly frontier cow-towns was quite remarkable. For example, one of the wildest of them all, Newton, was quickly tamed by the combined forces of the Santa Fe, *The Newton Kansan* (beginning publication in 1872), and the Temperance League; a local reporter boasted of the progress achieved by August, 1873, just before a small delegation of Russian-German Mennonites toured the area under the guidance and care of railroad agents."[12]

The AT&SF developed aggressive campaigns to get easterners and immigrants to move to the state. One of the largest ever marketing campaigns was launched.

Carl Bernard Schmidt - (1843 to about 1921) - Agent for the Atchison, Topeka and Santa Fe Railroad - The most significant person responsible for 1870s immigration of Russians to Kansas, without him neither I nor the threshing stones may have ended up in Kansas. KSHS

Stone #41, Private, Canon City, CO

Photo - Alan Ediger

there was a threshing stone in town that was owned by a teacher who grew up in the Buhler/Inman area with the last name of Ediger.

He later contacted me and said that it was Alan Ediger and that he now lived in Canon City, Colorado. I knew Alan when we were both young at Hoffnungsau Church. I contacted him and he told me his story of discovery.

When he was a boy, he was out in their pasture just west of Hoffnungsau and saw an odd-shaped stone in the washout. He told his dad about it and after some convincing got him to come check it out. They later got the tractor and dug out a threshing stone. He has been lugging it around the country ever since, and proudly displays an American flag above it using the stone as a flagpole holder.

Threshing Stone #42

This stone almost slipped past my radar, but someone said in passing that there may be a threshing stone at the museum in Halstead.

One day while driving through Halstead I found the threshing stone setting in front of the Halstead History Museum. I finally made contact with Caroline Williams, Museum Director, who told me that she and her brother Richard Basore had donated the threshing stone to the Museum. It had been in their father's possession, but

Stone #42, Halstead Museum

Newton Train Station - 1871. Harvey County Historical Society

The AT&SF company hired an implement dealer from Lawrence, Kansas to lead the campaign, Carl B. Schmidt. Without a doubt, C. B. Schmidt is the name that most often surfaces in the research of this book. I think he alone is the major reason the Russian Mennonites ended up in Kansas and without him the threshing stones would not be here either.

Cornelius Jansen, a Mennonite who was a Prussian consular official, read an article in the *Frankfurter Zeitung*, written by Schmidt who had corresponded abundantly with German newspapers, praising Kansas opportunities. Jansen wrote to Schmidt and asked if Kansas had enough room for several thousand settlers if they chose to come. Schmidt, peering over the letter at the vast stretches of virgin soil, chuckled and replied that it could accommodate several hundred thousand.[13]

To railroad officials, the news of a possible mass migration of skilled farmers from Russia seemed like manna straight from heaven. Because Schmidt had already corresponded with the Mennonites and German was his native tongue, the Santa Fe made him its Commissioner of Immigration in 1873.[14]

The land commissioners wanted people who would stay on the land and make it productive and discouraged anybody who would not be likely to stay with the task. The industrious, successful farmers of Dutch ancestry in southern Russia measured up to the request and were personally invited by representatives of the railroad.[15]

The land department of the railroad, knowing of conditions in southern Russia, was extremely anxious to turn the tide of this desirable immigration toward Kansas. To this end, Schmidt, was sent as a missionary to open the eyes of the Mennonites to the desirability of Kansas as a place for home-making and for the successful carrying on of the pursuits which had brought them prosperity in their adopted land.[16]

The Mennonite farmers, who only a century before had broken the soil and tamed the land in the Ukraine, were a perfect fit to do the same again on the plains of Kansas. These people were excellent farmers, had tremendous agricultural skills, needed more land, and were looking for religious freedoms. They had done it before; they could do it again.

The sales pitch to come to Kansas was strong. Schmidt made many trips to the Ukraine to convince the elders of the communities of the advantages of Kansas. The climate was similar, the terrain was similar, with rich black soil, no trees to clear, access to good water, and land was plentiful and cheap.

The Mennonites were not just looking for economic expansion, but with the threat of losing some of their religious privileges and possibly land rights, the opportunity to move was a great temptation.

To make the deal even stronger, the State Legislature amended the Militia Law of 1868 so that all people who, on or before the 1st day of May in each year, filed with the clerk of their county an affidavit that they are members of a religious organization, whose articles of faith prohibit the bearing of arms, shall be exempt from militia duty. This gave the Mennonite immigrants confidence that they could continue with their pacifist beliefs.[17]

But the biggest advantage was access to land. Additionally, the Homestead Act was put into place in 1862 by President Lincoln, which granted to U.S. citizens the title to 160 acres, provided they would live on it and cultivate the land for 5 years. This act was in particular to attract settlers to the wide open prairies of the Midwest.

The opportunities were great. The AT&SF needed to make this work to have any chance of surviving; the rail-line could not support the company alone.

The deal got even sweeter as time

they do not know the history of the stone.

At the time I photographed it I didn't know why one end was hollowed out. Later I realized this was a stone that had been re-purposed with the end hollowed out so it could be used as a salt lick stand for cattle.

Threshing Stone #43

This stone was a surprising find. I spent 2 days at the Goessel Threshing Days event, finding stone leads and having great conversations with lots of people. A person came up and said, "Did you know there was a threshing stone way back behind the steam tractors in the hedge row?" I didn't know of this one, but upon inspection it turned out to be a concrete stone used in past demonstrations and not an original.

While cleaning up my display at the end of the day, someone came by and asked if I knew about the threshing stone, over in the flower bed about 50 yards away from where I was sitting. I assumed this was another concrete one, but Kelly Harms and Johnny Schroeder (the two people that have helped me find more stones than anyone else) walked over just across the parking lot, and found that there was a real threshing stone in the flower bed in someone's back yard. If it had been a snake it would have bit me. So close yet almost never found.

Stone #43, Private, Goessel

Later I made contact with Irvin Goertzen the owner of the stone. This stone is from the J.H. Goertzen farm 3 miles east of Goessel where it was found out under the hedge row. They displayed it on its side in the back yard of their home. They have since moved the stone to another location.

Dietrich Gaeddert Application for Land, October 20, 1874 - from the Atchison, Topeka & Santa Fe Railroad Co. for 160 Acres @ $1.65. MLA - Dietrich Gaeddert archives

went on, as the immigrant farmers were able to negotiate an even better price for the land. The going price for railroad land was about $7.50 an acre, but the shrewd negotiations of the Mennonite representatives got them an even better deal, buying land for about $2.50 to $5.00 an acre.

The 15 October 1874 *Topeka Commonwealth* reported: "One of the largest bona fide land sales ever made in Kansas, perhaps in America, has just been concluded by the Atchison, Topeka and Santa Fe Railroad Company with a community of Russian Mennonites who landed in New York during the month of September, with the steamers Cimbria, Teutonia, and City of Richmond, and most of whom have spent the last thirty days and a good many of their rubles in our city. Their land purchase amounts, in round numbers, to about 100 thousand acres of railroad land, aside from a number of improved farms, all lying north of the sections of Florence, Peabody, Walton, Newton, Halstead, Burrton, and Hutchinson...from the Cottonwood River to the Little Arkansas River, a scope of magnificent prairie country fifty miles in length, is now one colony, composed of the thriftiest and most intelligent class of foreigners that ever landed upon our shores; and 'in three years'—to use the language of one of their elders—'that ocean of grass will be transformed into an ocean of waving fields of grain, just as

Immigrant ship the Teutonia - Author's ancestors traveled on this ship to the United States. Heritage Ships of the Norway, Heritage Collection

Mennonite dress - "Frank Leslie's Illustrated Newspaper," New York, 1875. MLA

we left our Molotschna colony.' Kansas will be to America what the country of the Black Sea of Azov is now to Europe—her wheat field."[18]

Additionally, even further negotiations were made. Special passenger rates were granted (prominent leaders even traveled for free) as well as free transportation on larger than normal amounts of goods. Eventually some negotiated with the company to charter a Red Star ocean steamer, which was sent to the Black Sea for a ship load of Mennonite household goods and farm implements. These goods were brought to New York and then by rail to Kansas, all free of charge to the colonists.[19]

Some writings stated that among the shipment was 100 of the Mennonite threshing stones. [See Article 5 that tells more about this story.]

It has been said that the Santa Fe made no profit on its sale of land to the Mennonites because of free services and

Threshing Stone #44

I first heard of this stone from a man who had a lot of information about antiques and threshing stones.

He said that there may be one on a farm east of North Newton on 24th. Also at a later date, Kelly Harms told me that he thought that there was one on a farmstead on east 24th. I stopped by the farm and no one was home, but I could easily see the threshing stone in the back yard, used as a bird bath stand. I stopped by at least 4 more times before I found someone home.

The person living there was renting the place and said the stone belonged to his landlord.

Stone #44, Private, Rural Newton

Threshing Stone #45

While at the Goessel Threshing Days, Terry Jantzen came to my table and said that he had a threshing stone, and

Stone #45, Private, Rural Moundridge

his brother had one too. His has always been on the farm that he now lives on. It had belonged to his grandfather, Jacob Strausz, but he is not sure if it was used by him or if he purchased it.

Threshing Stone #46

This stone is owned by Terry Jantzen. It is also from his grandfather Jacob Jantzen. His grandfather had a lot of antiques and he may have purchased this one along the way.

It is very typical except for the fact that it has 3"

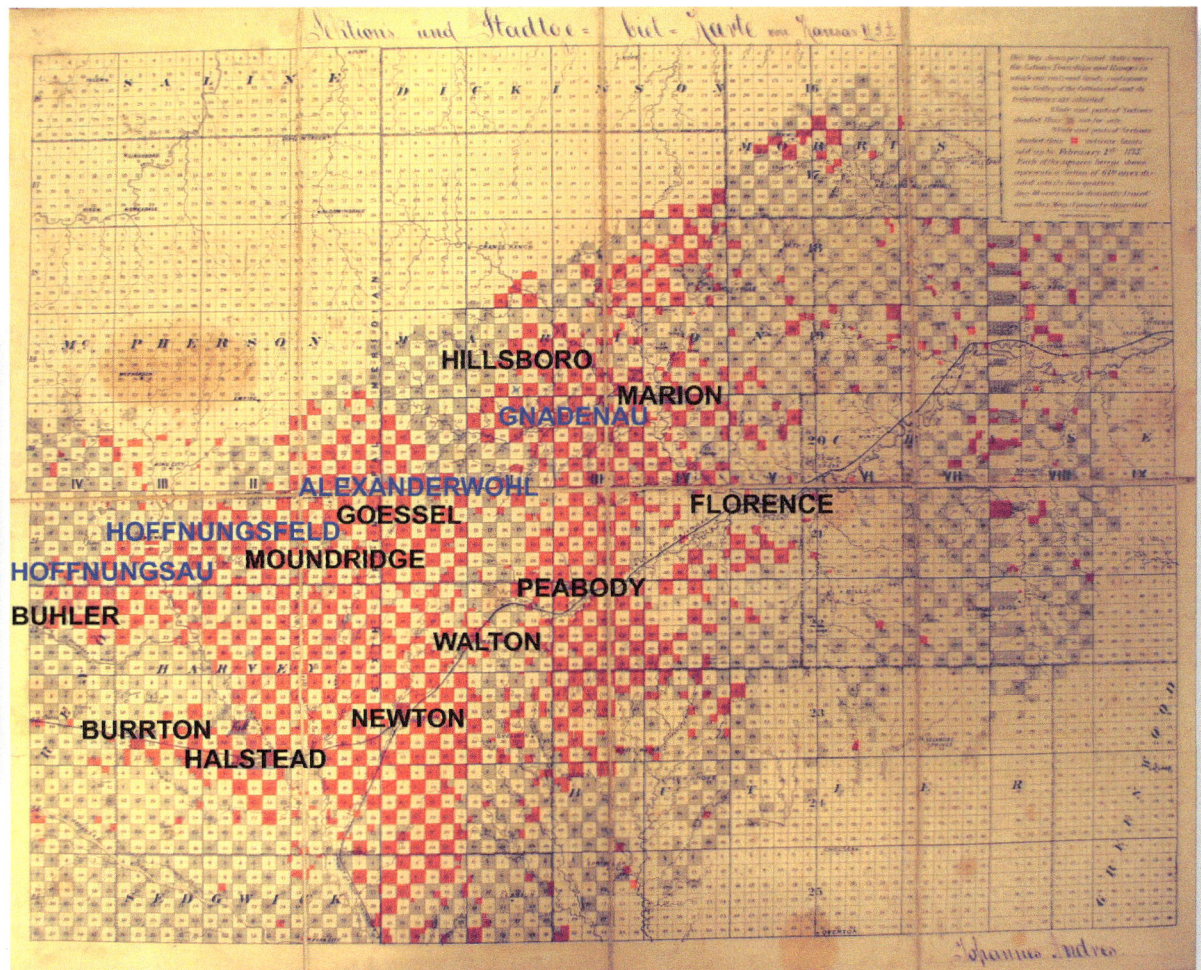

Map of Kansas,1875 -Shows the sections of land the Santa Fe railroad had for sale marked in gray and the sections marked in red were already sold to the immigrants, this is the exact same area in where the threshing stones were used. MLA

other general benevolence. The profit came later, after the hardworking thrifty farmers built up a heavy freight business for the railroad.[20]

"By the end of September, 1874, nearly 2,000 immigrants had arrived in Topeka, where the Santa Fe Railroad quartered them in a huge brick structure until the head of each family could select a tract of land and move onto it. Almost the entire populace of the Kansas capital turned out to stare and giggle at these strange, ludicrous creatures."[21]

The newspapers poked gentle fun at the foreigners. "The town," one paper said, "abounded with sheepskin coats, ample breeches, bulbous petticoats, iron teakettles, and other objects supposed to

Immigrant Houses - "Frank Leslie's Illustrated Newspaper," New York, 1875. MLA

be distinctly Russian." But almost over-night Topeka's attitude changed from derision to admiration and praise when its merchants suddenly discovered that these Mennonites had brought with them $2,250,000 in gold. Business began to prosper as the immigrants purchased vast amounts of farm implements, horses and cattle, and household goods. To show their appreciation for the sudden financial prosperity, local officials arranged a tour through the state capitol for all 2,000 people, to shake hands with the governor and observe the inner workings of the government.[22]

Upon arrival to central Kansas, the AT&SF even promised to build large immigrant homes. Two sections of land were given to them by the Santa Fe, and ultimately large immigrant houses, 18 feet by 200 feet were built as temporary quarters and housed approximately 400 people. These were built at several locations. The

Alexanderwohl immigrants, for example, lived during the winter of 1874-1875 in two large immigrant houses that were built 15 miles north of Newton near the present-day church building.[23] Another such immigrant house was located on Section 33 of Menno Township east of Hillsboro, another southeast of Inman near the Hoffnungsau Church, and one west of Moundridge near the Hopefield Church.

Interior of Mennonite immigrant house in Kansas - "Frank Leslie's Illustrated Newspaper," New York, 1875. MLA

square holes about 3" deep on each end and does not have a through axle hole. One can assume that a metal axle stub was wedged into this hole and that the short axle would spin in the towing frame.

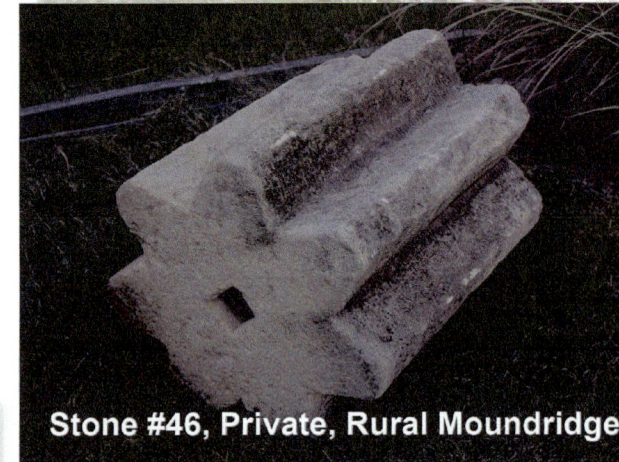

Stone #46, Private, Rural Moundridge

Threshing Stone #47

I received a phone call from Ethel Abrahams telling me that she and Norman had purchased a stone from a farmer near Hillsboro but unfortunately she could not remember the name. I went to take a picture, finding it beautifully displayed in the garden with a reconstructed wooden towing frame that had been made by her husband.

One curious thing about

this stone is that it has a small, precisely cut letter "O" in the end of the stone. This is unusual in that no markings were ever carved into any other stone that I have found. This remains a mystery as to who put it there and why.

Stone #47, Private, N' Newton

Threshing Stone #48

I got the lead to this stone from Ethel Abrahams. She said that Dr. Wilmer A. Harms had a threshing stone. After setting up an appointment I met with him and his wife. This stone was originally part of his family farm north of Goessel, the southern most farm in the Hochfeld Village. His family history is that his great

Another important contribution of the railroads was a supply of seed wheat for the first year.[24] Also during the first year lumber, flour, coal and household goods were shipped free by the Santa Fe to these new communities.

These families lived in close quarters through the winter or until private cabins, dugouts, lean-tos, and A-frames could be erected for individual families to start their farmsteads.

The economic impact on Kansas was significant. Not only did the railroad flourish, but local business also prospered with new commerce. One hardware store in Newton, for example, established a national retail record for the number of plows sold in one year, and some years later sold over 200 binders in a single season.[25]

Noble L. Prentis in *Kansas Miscellanies* wrote in 1889, "By far the most extensive and notable immigration in the history of Kansas was that of the so-called 'Russians,' which began in 1874, and which has resulted in the settlement of fifteen thousand Mennonites in the counties of Marion, Harvey, McPherson, Butler, and Reno."[26]

These farming families brought with them their farming skill, business expertise, tools, wheat, and millions of dollars in cash to change the plains from a wild, rich prairie to productive farm lands.

Immigrant House water well - "Frank Leslie's Illustrated Newspaper," New York, 1875. MLA

For the railroad it was a perfect plan. Unfortunately, it wasn't good for the native people or the buffalo. Turning the beautiful native grasslands into productive farmland would not have happened if it had not been for the ingenuity of the early railroaders.

ATCHISON, TOPEKA & SANTA FE RAIL-ROAD.

MAP OF KANSAS
ATCHISON, TOPEKA & SANTA-FE RAILROAD.

The Atchison, Topeka and Santa Fe Railroad Company are now offering for sale, in tracts to suit purchasers on **Eleven years** time, and at low prices **1,000,000** acres Superior Farming Lands, for Fruit Farms, Grain Farms, splendid Stock Farms and lands of every grade, at from $2.00 to $8.00 per acre.

For further particulars see the back of this map and address

D. L. LAKIN,
Land Commissioner, Topeka, Kas.

This road is now completed from *Atchison* via *Topeka, Emporia, Florence, Newton* to *Wichita*, and via *Hutchinson, Fort Zarah* to *Fort Larned*, being the only *All Rail Route* to South-western Kansas and the great valley of the Arkansas.

Lowest special rates made for Immigrants and their outfit. For further information address,

M. L. SARGENT, Gen. Freight & Ticket Agt. } TOPEKA, KANSAS. { T. J. PETER, General Manager.

References
County Towns are marked thus ✕
Villages .... thus ✕
Railroads Constructed
Railroads Building

The figures near Stations show the elevation to feet of the Depots above the ocean level.

Stage Lines

SCALE 18 MILES TO ONE INCH

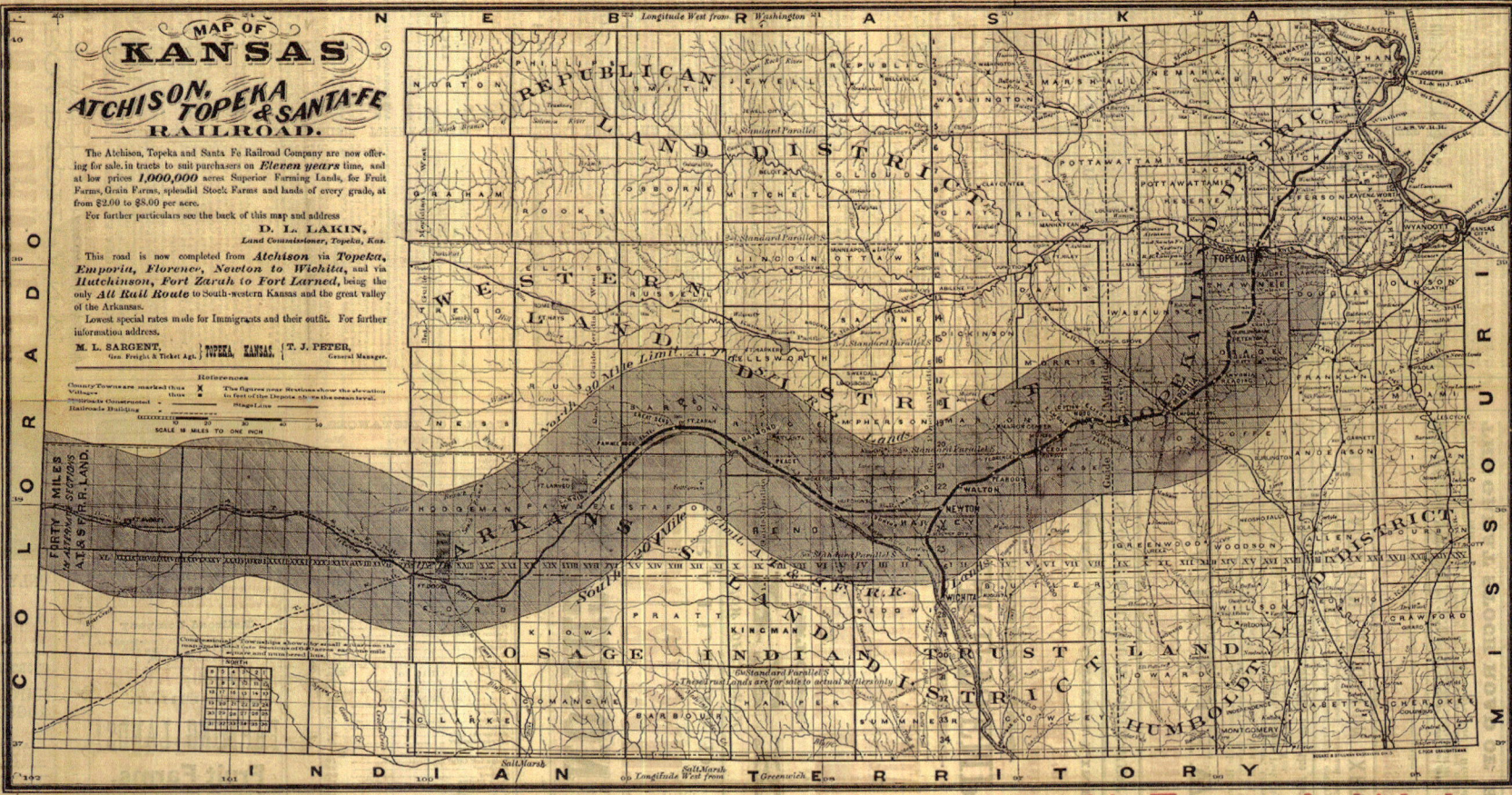

The Atchison, Topeka & Santa Fe R. R. Co. have 3,000,000 Acres of land in Kansas of which they are now offering 1,000,000 at from $2.00 to $8.00 per acre on eleven years time. 7 per cent. on deferred payment

Picking Kansas

What was so alluring about Kansas that in the 1870s thousands of immigrants picked Kansas as their new home? Of these immigrants there were as many as 16,000 Mennonites arriving on the plains to foster new opportunities and secure religious freedoms that were diminishing in the Russian homeland. By picking Kansas we end up with threshing stones virtually only in this state.

Today Kansas may struggle with some image problems. The state is not high on the tourism rankings, but in the 1870s Kansas was one of the most desirable locations in the world to embark on the new frontier and new opportunities.

What makes Kansas such a special state? James Malin of the University of Kansas and a prolific and extensive writer of Kansas history stated, "no field of historical study [is] more intriguing than the history of Kansas."[1]

The state was several times at the bottom of the sea, as layer upon layer formed the limestone bedrock of our state, capturing the bones of sea life and turning them into fossils replicating the ancient swimming creatures of a long gone era. The lands evolved into dry lands with bipeds and quadrupeds, carnivores and herbivores, roaming these same lands.

Mil Penner wrote about Kansas most eloquently in his book, *A Century on a Family Farm - Section 27*. "Section 27 lay as a plain of silt clay in which grasses and forests grew, died, decayed, and grew again. In the process, a thick rich layer of humus developed, anchored by woven masses of roots reaching deep in the soil. This luxuriant biomass supported and enriched an ever-more diverse cycle of plant and animal life, and the prairie became an immense storehouse of sun's energy."[2]

Time marched on, the area eventually evolved into open prairies, with forests confined mainly to streams. People first came to Kansas some 11,000 to 12,000 years ago during the last of the Ice Age. Although the state was not glaciated at that time, the climate was cooler and less seasonal than today. Huge animals such as mammoths and mastodons roamed the area until a gradual warming trend brought an end to the Ice Age and mass extinctions occurred.[3]

In the Paleoindian Period, from 11,000 to 7,000 B.C.E., the future state of Kansas was inhabited by nomadic hunters and gatherers, hunting big game with the use of spears tipped with large chipped stone points.[4]

During the Archaic Period the modern grassland environment of Kansas

Opposite page: 1874 Atchison, Topeka, & Santa Fe Rail Road Map of Kansas - Dark areas show Railroad properties. Wichita State University - Special Collections and University Archives

Arapaho camp with buffalo meat drying, Fort Dodge, Kansas
U.S. Department of the Interior - Office of Indian Affairs
1870. Wikimedia Commons

was established. The people of this time lived in fairly small bands as they roamed across large areas for food and shelter.[5]

The Woodland Period, A. D. 1-1,000 years, was marked by great changes in social systems. One of the most notable changes involved the widespread making of pottery vessels and chipped stone tools. Hunting tools became smaller, as the bow and arrow came into use. Towards the later part of the Woodland period, agriculture began with the cultivation of local plants and the introduction of plants such as corn.[6]

The Protohistoric Period, A.D. 1500-1800, refers to the time period shortly before and after the arrival of Europeans in the New World. Many of the Protohistoric Indian sites in Kansas can be identified with historically known tribes such as the Pawnee, Kansas, Wichita, and Apache. Some were nomadic and some

were involved in agriculture. The Apache lived by the hunt of the bison.[7]

The Historic Period began in 1541 with the arrival of Coronado. The French were next, some 200 hundred years later. The fur trade grew greatly during this period.[8]

The American presence began with the Louisiana Purchase in 1803. It is considered the greatest real estate deal in history. The United States purchased the Louisiana Territory from France, and effectively doubled the size of the United States and opened up the continent to westward expansion.[9]

Kansas became a territory in 1854 with the Kansas-Nebraska Act establishing the land covering the current state boundaries plus all the way west to the Rocky Mountain Continental Divide. The Act in effect repealed the Missouri Compromise of 1820 and allowed the new territories to decide if slavery would be legal there. Opponents of the Act helped found the Republican Party, which opposed the spread of slavery into the territories. As a result of the Kansas-Nebraska Act, the United States moved closer to Civil War.[10]

Abolitionist Free Staters from New England and pro-slavery settlers from neighboring Missouri rushed to the territory to determine if Kansas would become a free state or slave state. This created a hotbed of violence and chaos which was known as Bleeding Kansas.

grandfather went to a quarry north of Peabody for the stone. Then they carved on the stone all winter, working from wooden pattern.

The stone was to be sold at the farm sale, but Wilmer made arrangements to buy it and keep it in the family.

Stone #48, Private, N Newton

Threshing Stone #49

This stone was also a mystery stone for a while. I had three different people give me the lead to this stone. It was purchased by Bill Unruh at Ray Wiebe's auction. It is another one that has been reshaped for use as a salt lick stand. It is displayed in a flower bed in front of the house.

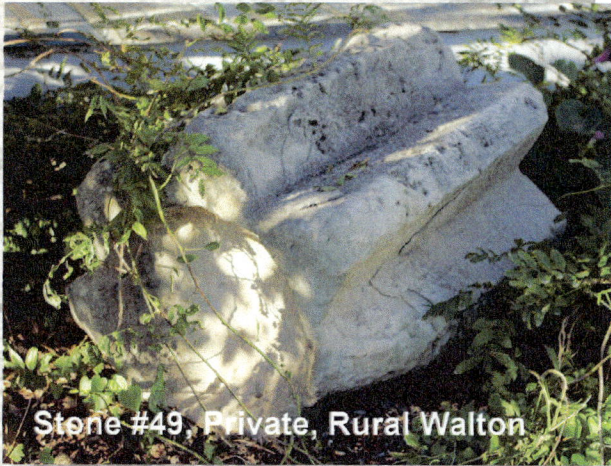

Stone #49, Private, Rural Walton

Threshing Stone #50

This stone eluded me for some time. Kelly Harms had heard of a threshing stone in west Wichita, Kansas. I found the address in the phone book, went to Google Street View and saw the stone in the front yard under a tree. I tried the phone number but it was not in service. So I drove to the address, where there was a for sale sign in the front yard, and the stone was gone, but I could still see the imprint of the stone in the ground-close, but not found yet. I had been told where they attended church, so I called the office to see if they attended there. The church office was very professional and said that they knew them and would give them my number.

The Lawrence Massacre, an attack on the pro-Union town of Lawrence, Kansas, August, 1863, due to the town's long support of abolition and its reputation as a center for Jayhawkers and Redlegs. Wikimedia Commons

The abolitionists eventually prevailed and on January 29, 1861 Kansas entered the Union as a free state.

But the state was not done evolving. The desire to bring commerce to the plains by the U.S. government and the State of Kansas was enormous. The belief in the potential of the rich soil and tempered climate envisioned limitless possibilities, but several things would need to change before this could happen.

Trails of travel traversed the Kansas plains. In 1822 the Santa Fe trade began, and the Santa Fe Trail, mostly following the course of the Arkansas River, became a feature of the region when surveyed and marked in 1825.[11]

At that time Kansas was nothing more than a territory to pass through, much like it is referred to today as a "Fly-over State." But this area was still the home of Indian civilizations, both nomadic and settled. If you had asked in the 1830s what the area was called, most would have said "Indian Territory."

The privileges of the immigrants were born on the backs of the Indians. The land was "claimed" by government institutional organizations outside of the Indian culture. The claim to their land was forced upon them. This is a sad and unjust component in the expansion of the West into Euro-centric organizations and government.

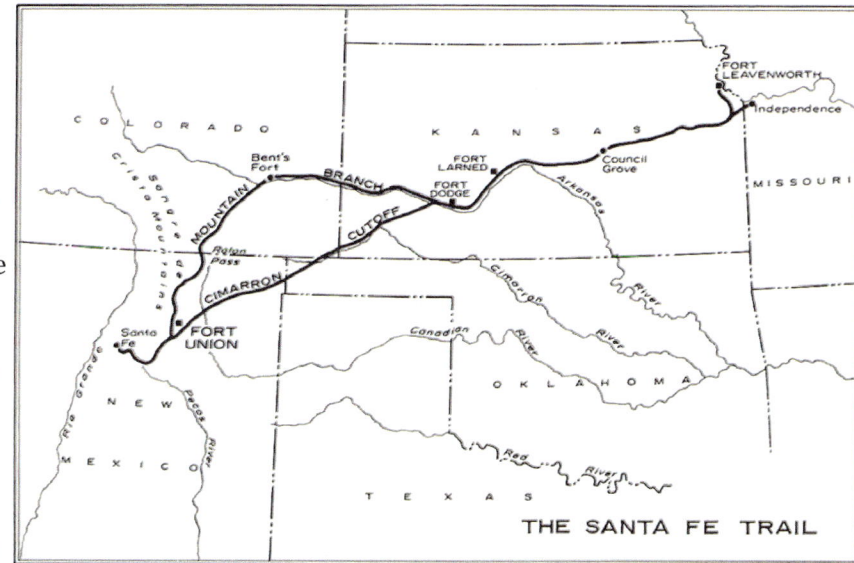

THE SANTA FE TRAIL

Map of the Historic Santa Fe Trail - around 1860. National Park Service, Historical Handbook Series No. 35, Washington, D.C. 1962, Public Domain

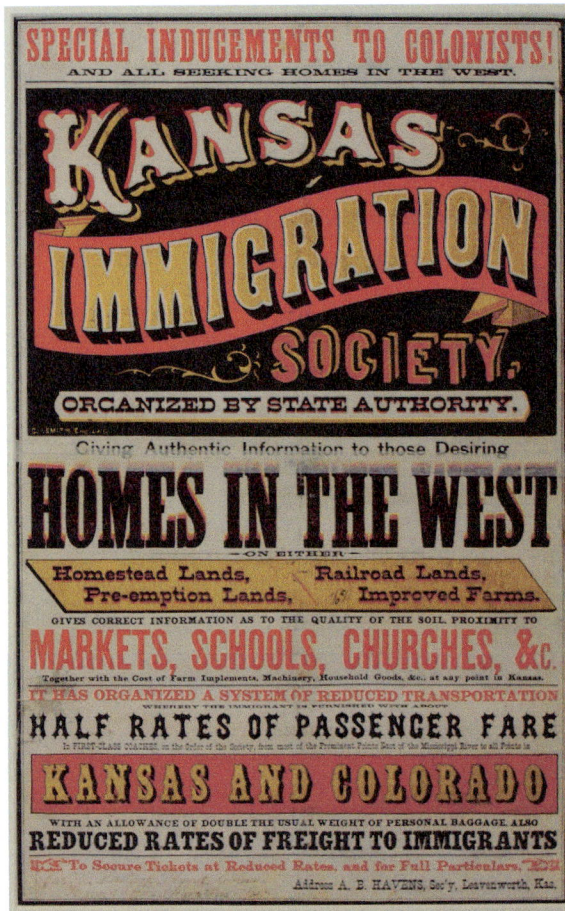

Kansas Immigration Society Poster. Kansas Historical Society Permanent display.

The relocation of Indians is also, a story of "removal" from Kansas. The permanent Indian preserves in Kansas, negotiated between tribes and the government, then ratified by Congress, lasted about a generation. When white settlement began its speedy growth with the founding of Kansas Territory in 1854, the Native Americans were pushed on again,

most often to Indian Territory (Oklahoma). By the end of the nineteenth century, most Native Americans were gone from the state or incorporated into its general population.[12]

The buffalo, a complete and total resource for the Native Americans, was soon to be eliminated too. The buffalo had become an obstacle to the development of the railroad like the Native Americans and they were all but eliminated from the scene.

This led way for the railroad to lay down track across the state. The richest land in the country was believed to be right through the center of Kansas. The Atchison, Topeka and Santa Fe Railroad dominated the development through the state roughly following the path of the Santa Fe Trail. Its lust for commerce provoked heavy marketing to farmers in the East as well as farmers from across the Atlantic. (See Article 8 about the history of the AT&SF.)

The marketing of Kansas to immigrants was also aggressively taken on by the State. This was a full attack to make Kansas look great to settlers. The State of Kansas published a booklet, *A Home for Immigrants - Agricultural, Mineral and Commercial Resources of the State - Great Inducements Offered to Persons Desiring Homes in a New County - The Homestead Law* (1863).

The handbook speaks flamboyantly about the State to induce all readers to come to this land of beauty and

Eventually they did call me, Gene and Jana Hildibrandt were living in Valley Center, Kansas. They gave me permission to photograph their stone. This stone has a crack running across the center of it but has been glued back together. It is nicely displayed in the front yard.

Stone #50, Private, Valley Center

This stone was originally on the Frank Wohlgemuth farm south and west of Hillsboro just west of the Gnadenau Church. It had been passed down to his oldest son and now is his grandson's.

I learned later that threshing stone #54 had also originated on the Wohlgemuth farm.

Threshing Stone #51

Several leads to this stone came to my attention at the Buhler Frolic, where I had set up a booth on Main Street to see if I could find any stones in the Buhler area.

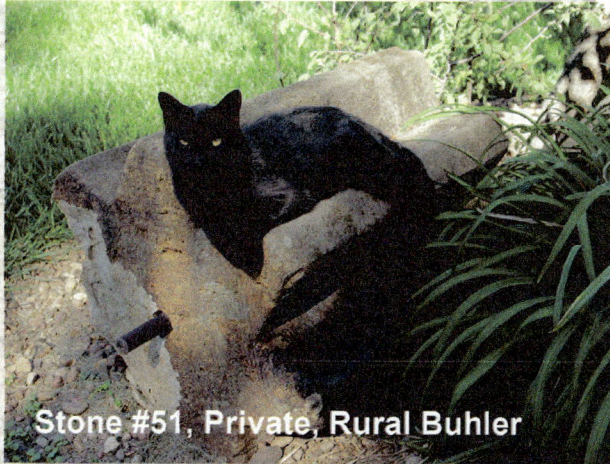

Stone #51, Private, Rural Buhler

This stone is owned by Gordon Schmidt. He had it displayed in his front yard with the cat lazily enjoying the cool temperature of the stone on a warm summer day. The stone is originally from the Goessel community.

Threshing Stone #52

This is the only threshing stone I have found truly by accident. I decided to check out

opportunity. "The climate of Kansas is, without exception, the most desirable in the United States...the winters are short, dry and pleasant, with but little rain or snow. The grass is green in the forests and on the prairies until mid-winter. And very often, herds of horses, mules and cattle roam at large during the entire winter. At the close of February we are reminded by a soft gentle breeze from the South, that winter is gone; and grand prairies, interspersed with every variety of flowers, and dotted by numerous herd of fine stock or perhaps a train of emigrants wending their way in search of new homes, assume their usual green robes of carpet, and present a scene of superb grandeur. During the summer there is always a cool, refreshing breeze, which makes even the hottest days and nights pleasant and delightful.

"The soil is deep, rich and fertile; in the valleys extending to the depth of four feet and resting on a clay subsoil; and upon the lowlands and broad prairies to the depth of from one to three feet resting on a subsoil composed of clay and sand. The richness of the soil is demonstrated by the luxuriant growth of prairie grass which is yearly produced."[13]

Who could resist the temptation to come to this "garden on the plains" to live and experience the wonders of this land called Kansas?

With both the State promotions and the AT&SF promotions they certainly got the word out.

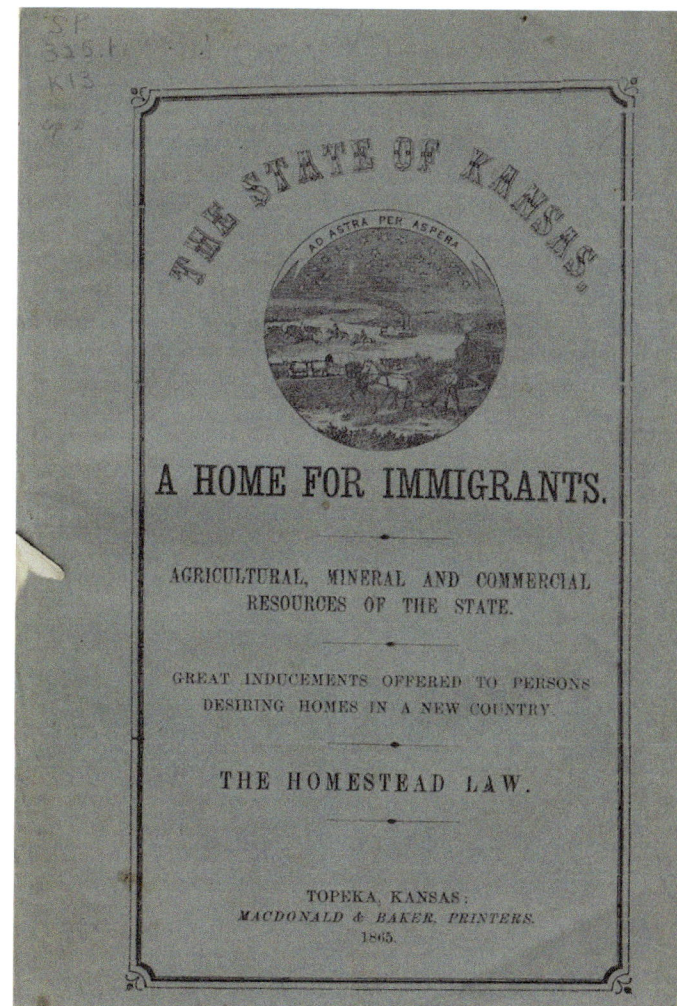

The State of Kansas - "A Home for Immigrants."
Kansas State Historical Society

C. B. Schmidt, an agent for the AT&SF, who spoke the German language and was aware of the plight of the Mennonite farmers in the Ukraine, made many trips to meet with community leaders to entice the Mennonites and other farmers from the Ukraine to Kansas.

The farmers in the Ukraine had run out of land, and religious privileges promised by Catherine the Great looked to be coming to an end. Leaders had already been looking at options for new territory to migrate to.

They considered many options starting in about 1870. Serious thought of large scale emigration from Russia had been considered by Cornelius Jansen when he realized that the privileges of the Russian-born Mennonites were disappearing rapidly. He inquired of fellow Mennonites in America and Canada about settlement prospects.[14]

In the summer of 1872 four young, wealthy men, Bernhard Warkentin, Jacob Boehr, Philip Wiebe and Peter Dyck left for America to visit Missouri, Kansas and Nebraska. All but Warkentin returned.[15]

Seeing that little could be done about their situation in Russia, the Mennonites chose a delegation of 12 men,

Three of the men who visited Kansas are shown in this group picture taken in New York City in 1872, seated left Bernhard Warkentin, seated right Peter Jacob Dyck, and. standing right Philip Wiebe. MLA - Warkentin

representing various communities, and sent them to America in May, 1873 to look for specific areas where new settlements could be located.[16]

"They came with a fourfold purpose: to locate cheap, fertile land; to obtain assistance in transportation; to determine whether they could enjoy religious freedom and be exempted from the military service; and to establish their right to live in closed communities with their own German schools and local self-government."[17]

The deputies spent 3 months in America exploring the open lands of Manitoba, Canada, and the Dakotas, Minnesota, Nebraska, Kansas, and Texas. Each man kept a diary and they met again in New York to compare notes and impressions.[18]

The reports of the 12 delegates were met with great enthusiasm and started the process for emigration to America.

Although the official delegates did not favor Kansas, Kansas became the preferred state of most

the McPherson Museum & Arts Foundation.

When I parked by the back door of the historic 1920s Vaniman Mansion, there set a threshing stone. It was displayed on its side next to the back door. The docent that day did not know anything about the stone, and later contact with the director confirmed that they did not know the provenance of this stone.

Stone #52, McPherson Museum

Terry Jantzen said that his grandfather had donated many things to the museum and that possibly he had donated a threshing stone to them too.

Threshing Stone #53

This stone came to my attention from individuals saying

Stone #53, Private, Inman

there was a threshing stone outside one of the entrances of the Pleasant View Home in Inman, Kansas. I easily found it setting upright on display near the greenhouse. This stone showed lots of deterioration similar to others that have been used as salt lick stands. The top end does not show a hollowed out area for a salt block but it is very possible that the bottom end does; this would explain its condition. The provenance of this stone is unknown.

Threshing Stone #54

This stone was first mentioned to me by Karen Penner who is on the board of American Historical Society of Germans from Russia in Lincoln, Nebraska.

It is on display on the grounds, shown with a wooden

of the immigrants. One reason why some Mennonites preferred Canada above any of the American states was the fact that they were offered large compact areas where they could live as they had in Russia.[19]

"Representatives who had an audience with President Grant found him a very kind and approachable man...but a person who was not able to make commitments to their requests. The matter was taken up in the Senate; the senator of Minnesota said that Canada was granting the Mennonites the rights for which they were asking, but that they preferred the United States. Since they were 'the very best farmers in Russia,' he was in favor of passing the bill. The senators of Vermont and Wisconsin were opposed to the bill, while senators from Indiana and Nebraska favored it. The latter said: 'If there is any portion of the world that can send us a few advocates of peace, in God's name, let us bid them welcome.' The bill was shelved."[20]

The Kansas legislature passed a law that exempted draft-age Mennonites from militia service if they registered

as conscientious objectors at the county courthouse.

However, this migration is not exclusively a Mennonite story. Along with wooing the Mennonites to the plains, other religion-based farming families were also enticed to migrate. However, the Mennonite groups tended to migrate in large masses, so this made them a very desirable group to invite.

By the time the first groups of Mennonites arrived in Kansas, representatives of the Volga Germans were inspecting land in the Untied States too. The trek of the Volga Germans was generally more difficult than that of the Mennonites, because they did not move in established church communities.[21]

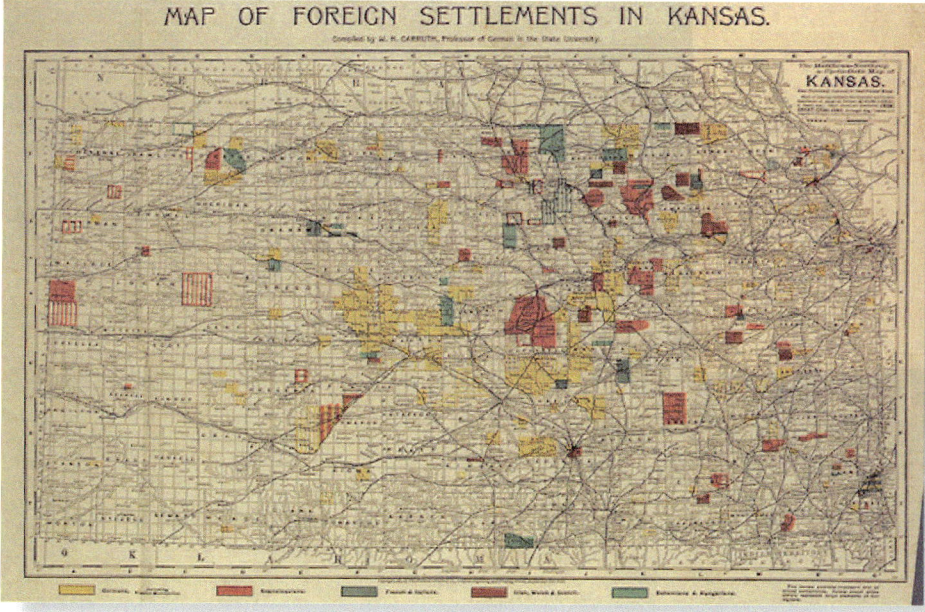

1894 Map with German and German/Mennonite in Yellow.
Kansas State Historical Society - William Carruth

The other major area of settlement in Kansas, in Ellis, Russell, and Rush counties, was colonized by the Volga Germans of Roman Catholic, Lutheran, and Baptist denominations. Of course, many counties of western and central Kansas became the homes of Russian-Germans, but many of these came later and often

Who would not pick the fertile soil of beautiful Kansas?

Stone #54, AHSGR, Lincoln NE

Photo courtesy AHSGR

towing frame made of large square lumber with steel reinforced corners, and a cross bar both in front of and behind the roller. The stone was donated to the organization and moved to Lincoln in 1996. It is originally from the Wohlgemuth family farm near Hillsboro, then it moved to the Herbert Wiebe family, and then Betty Koop donated it to AHSGR.

Threshing Stone #55

I set up a table display at the Buhler Frolic to collect information from individuals. While taking a lunch break I sat down with my cousin Dellis (Schmidt) Dyck and her husband Norman. After explaining about my threshing stone project, Nor-

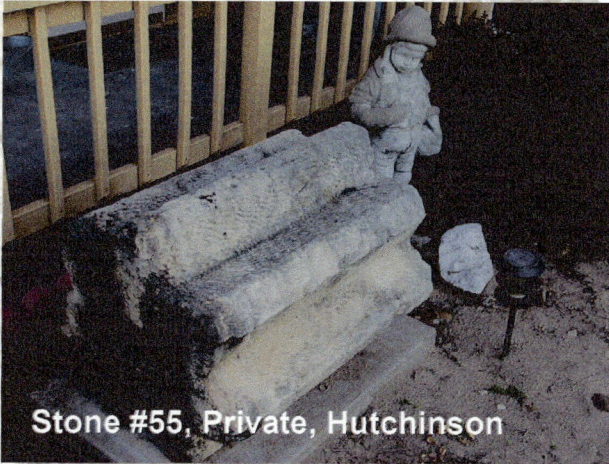

man said his dad had one and his brother Larry had it at his house in Hutchinson, Kansas.

I contacted Larry and took a picture of the stone, where he has it displayed in front of his house. The only history he knew, was that it was from the home farm of his father Frank Dyck.

Threshing Stone #56

I had heard of a person from Hesston who might have several stones. I later found out that his son was the person to contact, so I met with him and made arrangements to photograph the threshing stones.

We went out to his farm where he had 3 stones on display. They were all purchased by his father at auctions. One was

Buffalo are still being raised in western Harvey County.

involved people who immigrated first to other states or to Canada, Mexico, or South America.[22]

The majority waited until after the harvest of 1875 to travel to the United States. Not having the buying clout in numbers that the Mennonites had for leverage, made the trip much more difficult.

The other determining factor for picking Kansas was the climate. The Kansas climate was remarkably similar to that of the Ukraine, similar seasons and similar rainfall amounts, with slightly less harsh winters. The farming practices that had become highly evolved would transplant nicely in the similar climate of Kansas.

The soil and terrain were almost identical too. The loamy topsoils of Kansas closely resembled the soils of the steppes of the Ukraine, and were considered even better. Again farming practices could be directly applied to the new location.

The fertile soil, the climate, the terrain, the deals made by the railroads, the state government incentives and laws, the federal government laws, the need for immigrants, the desire for new opportunities all combined to make Kansas the destination of thousands of new United States citizens that now called Kansas home.

Prairie grass today in eastern Harvey County.

Farming with Horsepower

Harnessing the power of draft animals was a significant advancement in the productivity of humankind. The horses, oxen, mules, or camels could do the work of many men. But it also required year-round care and attention to keeping the animals healthy and productive. The entire farming year required the use of these highly valuable creatures.

"Horsepower" is a term we associate today with the measurement of the power that a machine can deliver. But for thousands of years it literally meant one horse equals one horsepower.

The use of "working animals" has been around for a long time. The power advantages that can be obtained from draft animals over humans is about 10-fold under continuous use.[1]

Farming with draft animals was certainly the method used by farmers 130 years ago both in the Ukraine and also on the plains of the United States. Following will be explanations of the use of these animals as the farmer went through the seasons involved in the growing of wheat.

A draft horse is a large horse bred for heavy work such as plowing, and other heavy towing tasks on the farm. Many draft horses could weigh up to 2,000 pounds; they have heavy bones and a muscular build. They tend to have broad, short backs with powerful hindquarters.

Work horse on a Mennonite Molotschna, Russia farm - circa 1900. MLA 2003-0288

They are intelligent and readily trainable to do hard work. However, the workhorse also was more expensive than the ox and required more attention even when not working.

Opposite page: 1902, German almanac called "Familien Kalendaer." MLA

"Although the horse was expensive to purchase, ate large amounts of costly grains, required elaborate harness, needed shoes for road travel, and was more prone to sickness than his counterparts, he was still the most desirable source of power. Several aspects that made the horse attractive as a power source were his speed, his ability to be commanded easily, and his proud nature."[2]

Much of the new draught farm machinery that was being manufactured during this period was designed specifically for use with the horse. The trend toward the use of horses was strengthened by the fact that horses could be used in a dual capacity, whereas the ox could only be used for field work and hauling freight. The ox was not socially acceptable for visiting, courting or church going.[3]

The ox was also used both in the Ukraine and in Kansas. Its characteristics were similar to the horse in that it was large, strong and had great endurance, but it was less expensive to buy and maintain.

Although the least desirable beast of burden, the ox was used to some extent during the 1850s as a power source. The animal worked at a very slow pace and lacked the ability to take verbal commands.[4]

With obvious disadvantages, the ox was still the most inexpensive source of animal power. The ox could be purchased for less money than a horse or mule. The ox usually lived a long life, was not susceptible to illness, and could be eaten when his usefulness was ended. The ox ate inexpensive grasses, required a simple inexpensive harness, and did not have to be shod for field work.[5]

There were advantages and disadvantages of either of these two most common hard workers. It was up to the owners to decide what would work best for them.

Team of 4 oxen plowing - circa 1920-1940. MLA - Farming

Farming in Russia with camels. MLA - Russia

Stone #56, Private, Rural Hesston

advertised for sale in the local newspaper and was on a farm west of Inman, another was purchased from Jerry Toews, and the third was from Harper, Kansas. This one was displayed in front of the house on a concrete pad.

Threshing Stone #57

Stone #57, Private, Rural Hesston

The next stone on the farm was displayed by a sidewalk on the other side of the house. It was displayed on its side. The end had been hollowed out for use as a salt lick stand.

Threshing Stone #58

Stone #58, Private, Rural Hesston

The last stone on this farm was displayed in the yard on its side and is in excellent condition, displayed setting next to an old implement wheel.

Threshing Stone #59

Numerous people told me to contact Virgil Litke of rural Hillsboro, saying that he was a great historian, that he most likely had a stone, and most likely knew other individuals that

Mules, the offspring of a male donkey and a female horse, are stronger than donkeys and were also used, but not to a great extent on the Mennonite farms. "The mule was tough, possessed stamina, and could outwork a horse. Generally, a mule was seldom sick, ate less than a horse, and did not require shoes. However, the mule did require the same harness as the horse and was often more difficult to handle."[6] Even camels were sometimes used on Mennonite farms in Russia.

The farming season began with the tilling of the ground. The plow or *mold board* (if it had a disk in front of the share) would cut about 3-4 inches into the ground and flip it over as it slid through the dirt, dumping it into the previous furrow. This was done annually, after harvest to renew the soil to a looser pack and to turn under the remain-

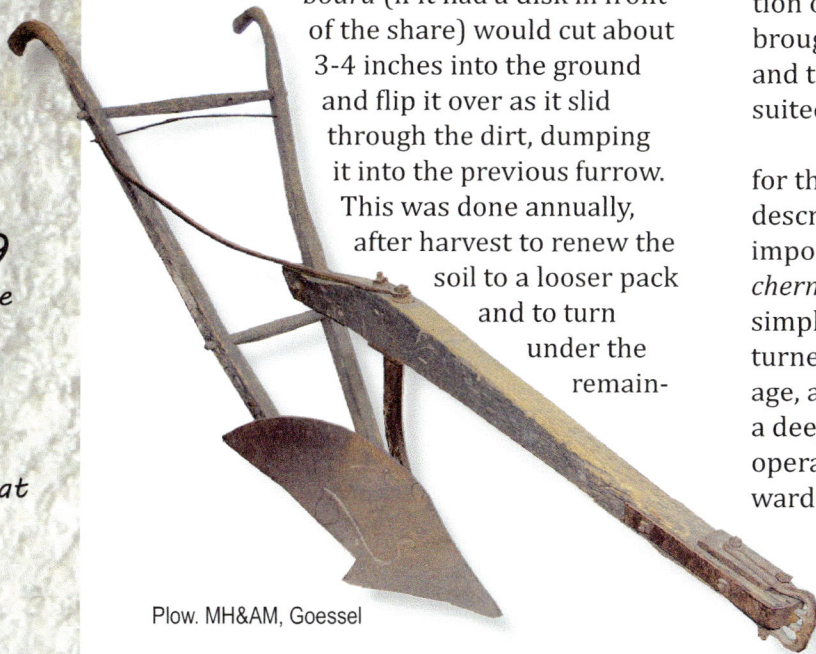

Plow. MH&AM, Goessel

Plowing in the Ukraine - circa 1900. MLA - Russian Farming

ing straw and stubble from the harvest, in order to allow it to decompose and create humus as it decayed under the ground.

Deep plowing was a concept used by Mennonite farmers in Russia due to lack of sufficient rainfall. "The Mennonites' contribution to the Kansas wheat industry was not limited to the introduction of hard red winter wheat. They also brought new techniques of deep plowing and thorough surface cultivation well suited to their new locale."[7]

Mark Carleton, in a 1914 report for the U.S. Department of Agriculture, described the tilling as follows. "An important feature of cultivation is the *chernui par*, or black fallow, called black simply from the very dark color of the rich turned-over soil. It is really summer tillage, and there are four cultivations. First, a deep plowing, and then three lighter operations at intervals of a month afterwards, made by very small gang plows or cultivators (chisels). Therefore, these people, on coming to the Great Plains, were already acquainted with the practices of early deep plowing and thorough surface cultivation, which…were widely advocated in this country in connection with dry farming."[8]

Plowing was the most demanding work for both the horses and the farmers. Plowing was done in the heat of the summer and was stressful on the animals, so extra attention was needed to ensure the animals received adequate water throughout the day. Regular rests were given to sustain their healthy performance.

The heavy work of plowing required the best equipment too. The collar, the leather covered padded ring that fits over the neck of the horse, was used to transfer all the lugging energy from the horse to the plow. The fit was critical, for the comfort of the horse and to obtain the best transfer of energy as it pressed against the shoulder bones of the horse.

The plow was attached to the collar via leather straps, or *trace,* that

Horse Collar. Author's Collection

continue back to a cross bar, or *single tree.*

For plowing and other heavy work, when more than one horse was used, horses could be ganged together to provide multiple horsepower. To accomplish this, a *multi-hitch* setup was required to be used. This consisted of the use of a *double tree*, for two horses, and an *evener* for four horses, these beams provided the equalized transfers of power from each animal to the load.

Usually the horse was outfitted with a bridle for control with reins. The reins were often slung around the back and over one shoulder of the operator, leaving hands free to control the plow, but the team was primarily controlled with verbal commands. The verbal commands used seemed to transcend language barriers; calling "Gee" turns the team right, calling "Haw" turns the team left even in Low German according to an interview with George Becker.[9] As seen in many threshing stone photos, a young operator could also ride the horse too.

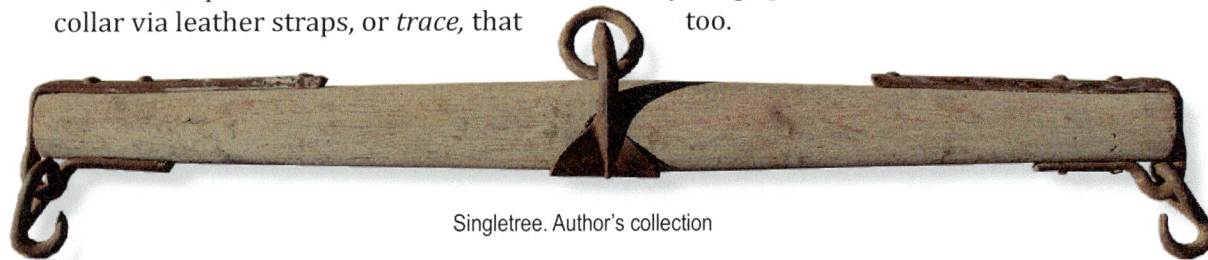

Singletree. Author's collection

might have a stone. I gave Virgil a call and he invited me out to his farm in the middle of a busy farm day. But he said "taking time to talk about history is more important than farm work and that I was very welcome to come out and we would talk about history." Virgil is truly a dedicated historian.

Stone #59, Private, Rural Hillsboro

At the farm he invited me to what he called the "Museum." I was very impressed with his collection of farming artifacts and his seemingly endless knowledge of the history of his collections.

He did indeed have a threshing stone, displayed in his immaculately kept farm museum. The threshing stone was

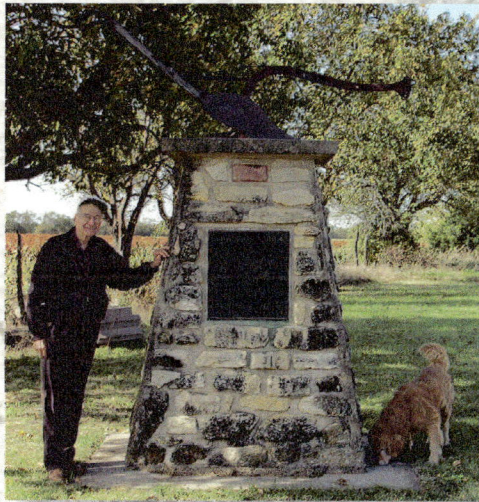

Virgil Litke by Trail Monument on his farm

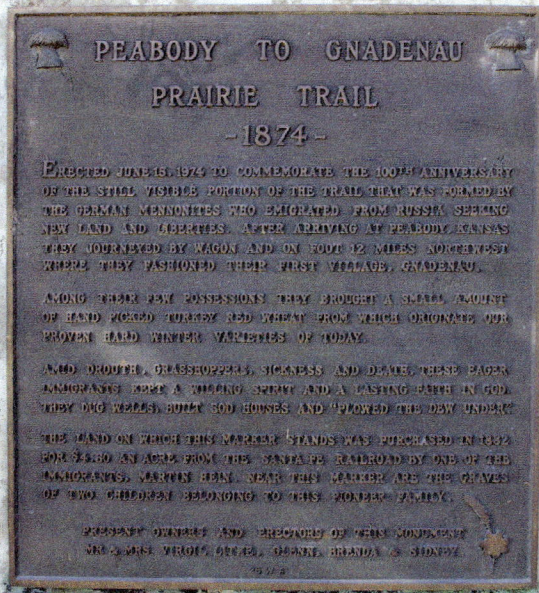

Standard Team Harness. Drawing by author - Source reference www.stitchnhitch.com

BROW BAND
CROWN
BLIND
THROAT LATCH
HAME
CHECK REIN
HAME BALL
CHECK LINES
NOSE BAND
BACK PAD
BACK STRAP
BIT STRAP
HIP STRAPS
BRITCHEN SEAT
BIT
COLLAR
HAME STRAP
COMBINATION SNAP
HAME CONNECTOR
SAFE
POLE STRAP
BILLET
BELLY BAND
QUARTER STRAP
LAZY STRAP
TRACE
HEEL CHAIN

displayed indoors on its side and had the end hollowed out for a salt lick stand.

Virgil has dedicated part of his farmstead to preserving the history that passed through his property. He has erected a

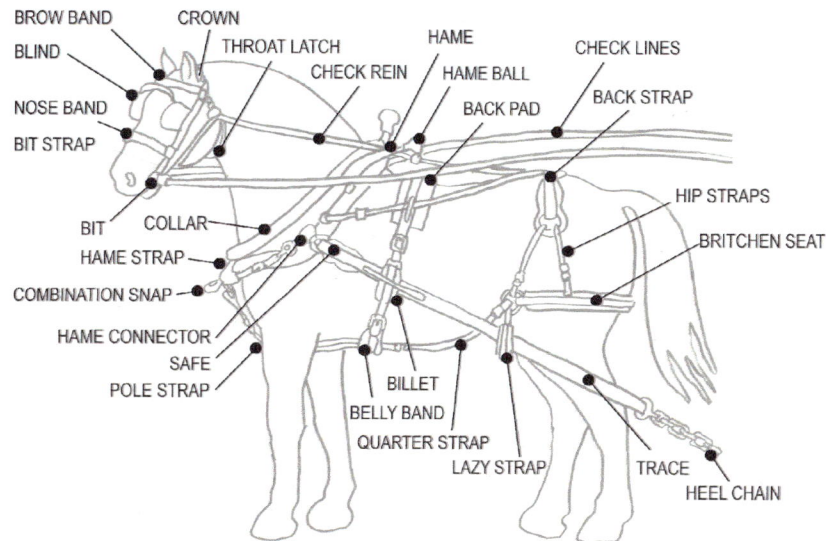

PEABODY TO GNADENAU PRAIRIE TRAIL

-1874-

Primarily male horses were used as draft animals in that they are stronger and have more endurance than the female horses.

Diligent training was required to create a team that functioned well together. This training was often done by a horse training specialist, but could have also been done by the individual farmer himself.

Other gear used for the horse sometimes included "blinders." These leather patches were positioned on either side of the head of the horse and held by leather head gear. Horses tend to be curious about their surroundings. To keep the horse from being distracted from their surroundings and focused on what lies ahead, blinders were used to focus their attention forward.

Following the entire team, harness, and plow was the farmer, or team operator. Virtually always a strong and mature man with the experience and endurance to walk behind the team all day, he gave commands and attention to the team.

The team could work all day and be rested and ready for the next day in the morning. Even with all the complexity of the harness rigging, an accomplished handler could rig a team of two to the plow in as little as ten minutes in the morning.

The long plowing day would start early. Tradition said, "We plow the dew under in the morning, and do not stop plowing till the dew falls at night."[10]

Henry B. Tiessen recalls the fol-

Plowing with 2-bottom plow with seat. MH&AM, Goessel

lowing in his book about the Molotschna Colony. "Early in the morning Father sounded the alarm and we would all get up. As a youngster I did not like this too well...but a farmer's boy had to get used to it and there was no way out. After breakfast we hitched four horses to the wagon and tied two alongside. Having arrived at the field...we hitched four horses to the plow and father would assume the reins and make the first couple rounds. He was quite expert at making the first straight line. He was very particular in making a

Faspa in the field, a late afternoon meal. MLA - Farming

straight and good beginning. With father having made a good beginning, the hired man would take charge of the plowing outfit while dad and I took the big forks and collected the weeds and put them in large piles.

"The biggest occasion was always lunch time. At half past 11 the big trough wagon was emptied and a feed consisting of hay, chaff, chop and oats was prepared for the horses. Water was added to the mixture and then mixed thoroughly. When being unhitched, there was always a great temptation by the horses to lie down and roll on the freshly-plowed ground. Before the feeding, they were watered and some drank several pails of it. In the meantime we made preparations for our own lunch. By means of tarpaulin some sort of lean-to was set up at the end of the wagon to protect us from the wind, and blankets were spread on the ground. We then took out our food, the casseroles, the borscht, various pastries, and big jugs of milk and coffee. Hard work and fresh air had generated a good appetite. After the meal the horses' crib was replenished again and we had an additional three-quarter of an hour of relaxation. Sometimes it happened that the neighbors or friends were working on their field nearby and I would skip over quickly and pay them a visit.

"All too soon the recess was over and the horses were hitched to the plow again. At this time the two extra horses were hitched to the harrow and it became my job to harrow the freshly-seeded ground. At four o'clock there was another little break when the horses were watered and fed, this was known as *Faspa* time."[11]

This tradition of Faspa continues on in many Mennonite families even today. In our family it was typical to have faspa brought to the field by my mother for the family to eat during harvest time.

monument on the path of the historic trail that passed through his farm. The monument is topped with a single share steel plow.

We talked for several hours, and he has continued to share information that has been interesting and helpful.

Threshing Stone #60

I heard stories of a stone in east Wichita from Kelly Harms, Leann Toews, and Johnny Schroeder. All had remembered seeing a stone in east Wichita on a main east/west road. I drove up and down East 1st, 2nd, Douglas, Central and 13th streets several times looking for a stone but never spotted one. I even called Tom Harder, pastor at Lorraine Avenue Mennonite Church, in east Wichita, to see if he knew of any church members that may have a stone, and he did not. I thought I had hit a dead end.

Months later, while interviewing Ray Wiebe, he was able to possibly solve the mystery. This stone #60 may be the stone of East Wichita.

Ray said Cornelius Duerksen immigrated to the U.S. in the late 1800s and was a stone mason. He had brought with him from Russia a wooden pattern. He reportedly had carved one or more threshing stones here in Kansas, even though they would have been made too late to ever have been used. His son George Duerksen lived in east Wichita.

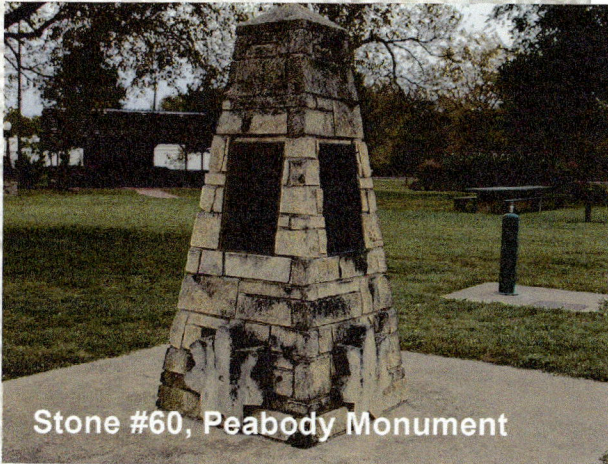
Stone #60, Peabody Monument

Ray remembers that the Duerksen stone was delivered from Wichita to the Hillsboro museum. At this point we don't

Wooden ox bow. Kauffman Museum

Of course many farmers also used oxen. This beast of burden used different rigging. The oxen transferred their pulling power to the plow with the use of an "ox bow" collar.

This is usually a wooden device made of a large beam that is contoured to fit over the necks of two oxen. Holes in the beam allow 2 "U" collars to fit around the neck of the ox and up into the cross piece. This collar also pressed against the front shoulder of the oxen to transfer the pulling power to the plow. One cannot imagine that the ox collar was very comfortable for the ox, but it has a long history of use from all around the world.

Following plowing in later summer the plowed soil, which has large clods, needed repeated rains and time to start mellowing, so the clods could be broken up. It was always critical to plow with the ground at the right moisture content; too wet, and the clods are large and become very hard; too dry, and the ground may be so hard that it is difficult to pull the plow through the dirt. However the advantage of the newly plowed ground is that the rough surface would capture

more of the rare summer rainwater.

Ideal soil conditions consist of *loamy* soil, which includes the right balance of sand, silt, and clay in relatively even concentrations of about 40-40-20.[12] Loam soils generally contain more nutrients and humus than sandy soils, they have better drainage and infiltration of water and air than silty soils and are easier to till than clay soils. Too much clay and the soil is tough. Too much sand and the soil is prone to blowing and quick loss of moisture,

To restore and maintain quality soil conditions, attention to soil management is essential. Plowing under the straw from the last wheat crop helps promote good soil conditions. The primary purpose of plowing is to turn over the upper layer, burying remains of the previous crop allowing them to break down as well as to help aerate the soil.

Over the year much manure would collect in the barns and corrals. The manure however was not waste, but valuable fertilizer. Manure when adding it to soil contributes to soil fertility, by adding organic matter and nutrients back into the soil, creating a humus, a complex organism that results from the breakdown of plant material, which is important to the fertility of soils in both a physical and chemical sense.

The manure was collected by shovel and scooped into the manure spreader. This wooden wagon could carry the manure to the field, being towed by horse. Then a lever was engaged to transfer mechanical power from the wheels to a chain that dragged the manure out of the box to the back where spinning tines helped to spread the manure over a broad path. Manure spreading was not always fun but wonderful for getting rid of the odious waste and turning it into rich fertilizer for the wheat.

Wood frame harrow. MH&AM, Goessel

Manure spreader in Russia. MLA - Farming

In a 2012 interview on National Public Radio, author Gene Logsdon said that there was a myth that American soil was so rich that it did not need manure to replenish nutrients, but by the mid 1800s it became obvious that manurial fertilizers needed to be applied.[13] Kansas Mennonites were always doing this.

Over the winter and early spring the wheat fields in their young green grassy state made excellent grazing fodder for the cattle and other farm livestock. Animals would be allowed to graze the field, while making direct deposits of fertilizer onto the fields. An adult cow can produce about 45 lbs. of "fertilizer" a day.[14]

Later in summer the plowed field needed to be worked multiple times to break up the clods and start to create a softer and receptive seed bed. Weeds would also sprout and cover

Wooden V-frame harrow. MH&AM, Goessel

MENNONITE CENTENNIAL MEMORIAL
1874 – 1974
THIS MONUMENT IS AN EXPRESSION OF DEEP GRATITUDE
—TO THE CITY OF PEABODY FOR RECEIVING THE MENNONITE IMMIGRANTS, EXTENDING HOSPITALITY TO THEM, AND NOW GRANTING A PLOT FOR THIS MEMORIAL,
—TO THE SANTA FE RAILROAD FOR DIRECTING THE MENNONITE IMMIGRANTS TO A PRODUCTIVE LAND AND MAKING IT AVAILABLE UNDER FAVORABLE TERMS,
—TO OUR GOVERNMENT FOR RESPECTING CONSCIENCE, REWARDING INDUSTRY, AND GRANTING FREEDOM FOR CHRISTIAN OUTREACH,
—TO OUR GOD FOR HIS LEADING AND CONTINUED BLESSING.
ERECTED THIS 27TH DAY OF JULY, 1974, BY THE MENNONITE INTER-CHURCH CENTENNIAL COMMITTEE, HILLSBORO, KANSAS.

DURING THE 1870'S AND THE EARLY 1880'S ABOUT EIGHT THOUSAND MENNONITES MIGRATED FROM THE STEPPES OF SOUTH RUSSIA TO AMERICA, AND ARRIVED NEAR THIS SPOT IN PEABODY, KANSAS. BECAUSE OF THEIR DEDICATION TO GOD AND THEIR FAITH IN JESUS CHRIST, THEY SOUGHT A LAND WHERE THEIR DESIRE FOR RELIGIOUS FREEDOM WOULD BE HONORED.

AMERICA BECAME THAT LAND. THE TERMS OF THE SANTA FE RAILROAD AND THOSE OF THE GOVERNMENT FOR LAND PURCHASES WERE ATTRACTIVE, AND THE VAST, FERTILE PLAINS AND FAVORABLE CLIMATE MADE KANSAS THEIR CHOICE. THE RECORD OF THE PAST ONE HUNDRED YEARS REVEALS THAT THE MENNONITES MADE A WISE DECISION.

Boy with a 4-horse team pulling spring tooth harrow - circa 1920. MLA

the field; these too needed to be eliminated to help preserve soil moisture. To accomplish these tasks a very valuable implement called a *harrow* was used.

The harrow could be made of wood, steel or a combination of both. The harrow is essentially a grid of cross pieces with spikes on the bottom side. This lightweight device could be widened by attaching multiple sections together to create the desired width. The harrow could be pulled by one or more draft animals.

The harrowing process would need to be repeated several times, to continue to break down clods and also for weed control. Towards the end of summer the farmer would look for "volunteer" wheat to sprout up with the weeds. This was actually seed that had fallen to the ground during harvest and had voluntarily sprouted and started to grow. The volunteer, however, needed to be killed with the weeds, because it would sprout too early and would die in the winter. So killing it kept them from drawing away moisture from healthy plants.

Ultimately the ideal soil conditions would be created with the harrowing process, creating the ideal seed bed for planting, a light fluffy soil that had an even and smooth texture with the right moisture content about an inch down for the seed to sprout in.

Sowing was accomplished by different techniques depending on the equipment available. The most ancient of methods was "broadcast" seeding. This was merely grabbing handfuls of seed in your hand and tossing out the seed left to right in a broadcast pattern, as evenly as possible as you walked back and forth across the field. A slightly more precise method was a hand crank seed "spreader" that would mechanically broadcast the seed

Broadcast spreader. KM

Horse drawn drill. MH&AM, Goessel

from a hopper carried by the sower as he walked the field.

This process was followed by some form of raking in the seed, either by hand or most often done by another pass of the harrow over the scattered seed to help embed it in a light covering of fresh soil.

The most advanced method used both in the Ukraine and then later in the U.S. was the "drill." This mechanical device pulled by horses could plant a width of about 6 feet. The drill had a hopper to carry the seed. Mechanical disks at the bottom of the hopper dispensed seed into a flex tube that directed the kernel to directly behind a *shiv* or hoe that created a small

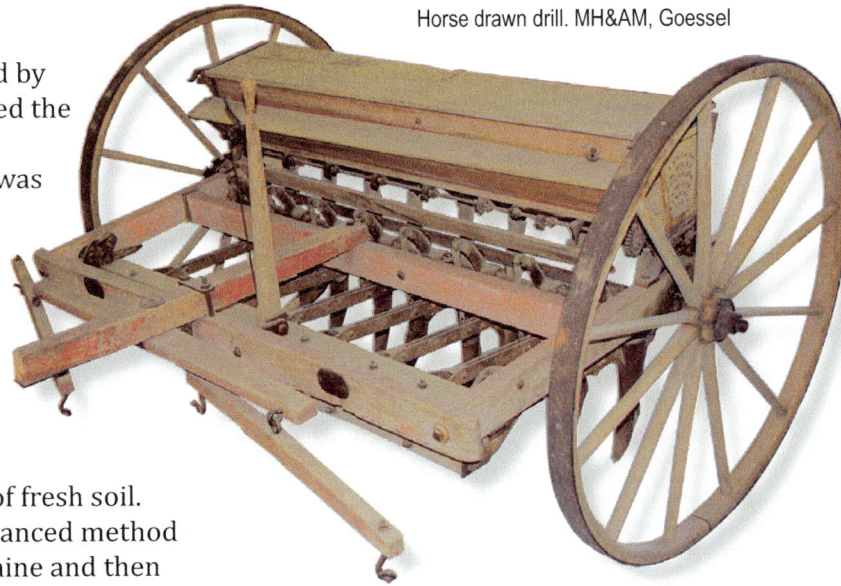
Horses pulling drill in Russia. MLA - Farming

furrow in the soil, allowing for the ideal placement of the seed a few inches into the seed bed.

Wheat sprouts usually a few weeks after planting, given the ideal conditions of rain and temperature. The hard winter wheat grows leaves or *tillers* until being interrupted by the onset of winter when it is too cold to grow. It then goes into a dormancy state for the winter. The blades start to yellow and shrink, and it starts to willow down for the long cold winter, but this is normal and part of the winter wheat life cycle.

The wheat loves to be under a blanket of snow, protecting it from drying out, and actually the snow insulates it from the extreme cold conditions. In late winter as spring starts to warm up the soil, the wheat plant begins to come out of dormancy, greening up and starting to grow again.

know for sure which one of the stones that one may be.

Then, Ray added, that this stone had once been painted red, white and blue, consistent with my earlier discovery about stone #19.

He said the Adobe Museum gave this stone to the City of Peabody for a monument, and that it had been cut into 4 pieces for that purpose.

I was shocked that someone would destroy a threshing stone by cutting it into pieces. As soon as our meeting was over I drove to Peabody to find the Mennonite Centennial Memorial in Santa Fe Park at the south end of Main Street.

My fears were immediately calmed when I saw how tastefully the stone had been used in the making of this monument, thus preserving it for a long time. The stone sets with one face on each of 4 sides protruding out of the base of the slightly tapered limestone masonry, with a brass plaque on each of 4 sides.

Threshing Stone #61

The location of this stone was known by Virgil Litke.

Much later I also received a letter from the owner, who had read about my project, letting me know of its location, on a farm a few miles southwest of Hillsboro where it sets upright next to a huge tree on the farmstead. The stone photographed beautifully on a nice autumn day.

Stone #61, Private, Rural Hillsboro

Threshing Stone #62

I got the lead to this stone from my website. A person had been surfing the net and ran across my project, read about it and realized that she knew where one was. She

McCormick reaper mower - circa 1830-40. MH&AM, Goessel

The next stage is the development of *nodes* becoming visible in the stem, followed by the *flag stage* when the last leaf begins to emerge from the *whorl*.[15] The critical spring rains are very helpful to encourage this growth.

The wheat fields by April are a beautiful carpet of rich dark green. By mid-May the heads starts to emerge from the tip of the plant, referred to as the *boot stage*. The green starts to change color slowly over a few weeks, ripening in June to a rich golden color, "the amber waves of grain." As the wheat approaches *dead ripe* the head will start to lean over and show us that it is finally harvest time.

However most often when wheat was threshed by hand the wheat was actually cut while still a bit green, but after the berries where fully developed in the head. This wheat would still ripen even though

Reaper -Stein der Weisen. A. Hartleben's Verlag - 1889. Wikimedia Commons

it had already been cut while in the shocks, but be dead ripe by threshing time.

Harvest time was also a time for horse power. Although many tasks were still done by hand, such as cutting the wheat with a sickle or scythe, the advancement of machinery and the use of horses greatly improved the productivity of the farmer.

Advancing to the mower technology for cutting the wheat was a great leap. Cyrus McCormick first sold his new invention in 1843. The *reaper mower* was pulled by a team of horses and powered

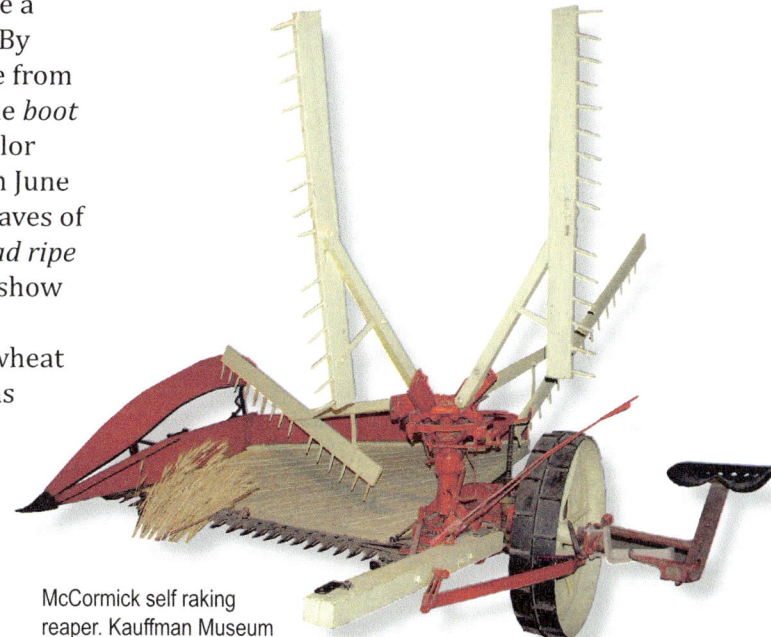

McCormick self raking reaper. Kauffman Museum

Community threshing using the large ladder wagons "Leiterwagen," Ohrloff, Molotschna. MLA - Russian Farming

by the drive wheel as it rolled across the ground transferring mechanical power to a *cycle blade* cutter, similar to a scissors that went back and forth between fixed teeth to cut the straw. The cut straw then landed on a platform where another operator would rake the straw off into piles that would later be hand bound into sheaves and then placed into shocks.

Mower technology continued to advance a bit more with the invention of a *self-raking side reaper* or *sail reaper* in 1862 that mechanically swept the straw to the side of the mower reducing the need for one operator and increasing output to five acres a day.

Horses were used to transport the sheaves from the field to the farmstead for threshing. In Russia the Mennonites used a special wagon called a ladder wagon for this process. These large long V-shaped wagons could carry tremendous amounts of cut wheat in one load.

Horses were obviously used in the threshing process, either for treading the grain or pulling the threshing stone. But the threshing stone itself was slowly being replaced by the mechanical stationary threshing machines. These devices, often placed in the yard, would be powered by 6 to 12 horses attached to a *horse engine* or *horse sweep power*. A rotary *capstan* allowed the horses to walk in a circle while each being attached to a radial *sweep pole* that transferred the mechanical power to the threshing machine. This same concept could also be used to pump water or mill grain.

sent me an e-mail saying that there was one in a pasture near Goddard on the west side of Wichita, but she did not know whose it was.

I drove there that evening and saw the stone sitting on a beautiful grassy knoll on an old but very pretty farmstead, but with no farm house. The barn was in the back yard of a house accessible from the other side in a large housing development.

Stone #62, Private, Goddard

I drove into the development to the house that appeared to be part of the same property. The woman was renting the place from her employer. She gave me his name and permission to photograph the

Horse engine for threshing machine power. MH&AM, Goessel

stone. I have tried repeatedly to contact the owner but still have not succeeded.

This stone is another one that has been hollowed out for use as a salt lick stand.

Threshing Stone #63

This stone came to my attention from a conversation with Don Huebert who thought Bill Hinz might have a stone. I contacted Bill and he invited me over to photograph his stone. He had bought his stone from Frank Pauls in the Buhler/Inman area.

Stone #63, Private, Rural Hutchinson

His stone was nicely displayed on the back patio. He recalled his grandpa Jacob Ediger saying that when he was a boy he rode the horse while

Cart Horse - Franz Martens family, Rückenau, Molotschna, Russia. MLA 2003-0275 - This photo was selected originally just for the horse and cart, then after further research it was discovered that the Martens family are ancestors of the author..

Work horses were brought to Russia from West Prussia but degenerated until the stallions obtained from the Don Cossaks and later Government studs improved the breeds. In 1855 there were about 10 horses per farm.[16]

Cart or *carriage* horses would also be used throughout the year for light duty, such as local trips to the store, hauling sacks of grain to market, or just to go to church on Sunday. For lighter loads a *breast collar harness* would be used, consisting of a padded strap running around the chest from side to side.

Year-round on the farm "Horse Power" needs care. It is a huge responsibility to care for these majestic animals that are essential to the productive farming process. Food, shelter, water, health care, grooming, and hoof maintenance, are all-year tasks. But the relationships developed between the farmer and these wonderful animals were also rewarding. Horses could work for up to 25 years, before being retired. In that amount of time relationships became strong and deep.

Fig. 26.

First Person Accounts

Finding first person accounts from individuals who lived through the early farming process and actually wrote about using the threshing stone has been one of my biggest quests. These people experienced it and told about it in their own words. The most amazing discovery was finding and interviewing George Becker who actually remembered using the threshing stone as a boy in Russia in 1919.

Following are stories in the words of those individuals who have written about their memories of farming and using the threshing stone. I truly am grateful to these individuals who wrote down their memories for us to enjoy. Read them with the understanding of the time about which they were written and the insights that they give, and for a better understanding of a time gone by.

ALEXANDER PETZHOLDT (1810-1889)

Report written in German by Petzholdt, Professor of Agriculture at the University of Dorpat, who on behalf of the Russian government made extensive exploration to the Mennonites of the Ukraine.

"The threshing roller - The Mennonites threshed the grain, as long as the favorable weather permits, in the open. The threshing is done with a stone roller whose surface is deeply grooved, drawn by two horses. It is similar to the ones from the area of Bachmut which weighed about 10-20 Pud; the ones used in the colonies have a weight of approximately 12 Pud [450 pounds]. During threshing, two people turn the threshed straw with wooden rakes, and then place the straw in large heaps, to be used for bedding and feed. The grain is brought together in the center of the threshed course on a heap, where more cleaning of the seed takes place with common fanning mills."

Alexander Petzholdt. Wikimedia Commons - University of Tartu Library

Alexander Petzholdt, "Reise im westlichen und südlichen europäischen Russland im Jahre 1855". Oldest known Illustration of a Mennonite threshing stone in the Ukraine. Illustrated by Russian scientist documenting the farming processes in 1855. MLA

MARIA (WALL) REGIER (1859-1947)

Maria's story was found in a book at the Harvey County Historical Society and also given to me by her grandson Ray Regier. Ray says, "I have a story told by my grandmother, Maria (Wall) Regier, as recorded by her son, my uncle, C. C. Regier, who wrote *Childhood Reminiscences of a Russian Mennonite Immigrant Mother 1859-1880* in 1941. Included in that story is the following."

"About the threshing of grain, too, it was different from what it is now. Generally we hauled all the grain home and stacked it. The stacks were arranged all around the threshing place. The place had to be prepared and hardened. The grain was then scattered and threshed with a threshing-stone. It worked very well, and I would know how to do it yet. Sometimes we got a hundred 'tschetwert' (over 500 bushels). I worked hard in the harvests, and it was much fun for me. I preferred it decidedly to household work. At such times father was always a little proud of me. I didn't know it then, but I know it now.

Maria (Wall) Regier.
Courtesy Ray Regier

"Harvesting was done with the scythe. One woman could bind what two men cut. I recall one occasion when mother had it all packed on a wagon (wagons were smaller than American wagons). Then, apparently, she turned about too abruptly. At any rate, the wagon tipped over and everything lay on the ground. What happened further I do not know, but I remember how mother's face flushed – just as Mary's did at times. Later, mowing machines were introduced. Once father went to Odessa and bought several. They were put together on our yard. Father had to do that as the other didn't know how. When I was big enough to help with the harvest the mowing machines were already in use."

GERHARD J. SCHARTNER (born 1898)

This memoir was brought to my attention by John Thiesen of the Mennonite Library and Archives. This text was forwarded to John from Alfred Neufeld in Paraguay. Schartner was born in Russia, ended up migrating to Paraguay in 1929, and wrote, *Lebensgeschichte und – Erinnerungen* (Life Story and Memories).

The memoir was translated by Ray Regier.

"When all the grain has been cut, then they take two of our work wagons apart on the yard, shove the rear wheels far enough back that we can put in a longer connecting pole to connect the front and back wheels. So then you have a wagon on which you can pile as much mowed grain as two strong horses can pull.

"When the grain in the field has dried enough, it is taken to the yard on such a wagon. Here the grain is unloaded onto the hard threshing floor. This loading threshing the wheat.

He had also recalled seeing two threshing stones in Dodge City at one time but they were gone now. No additional information about these stones has yet turned up.

Threshing Stone #64

I searched for this stone for months after getting my first lead from Charles Graber of Wichita. I work with Charles and he was talking to an acquaintance of his who told him that "Sprig" Graber had one in front of his home in Pretty Prairie, Kansas, at one time, but that he is no longer living and didn't know where the stone was now.

On the way to photographing another stone on the west side of Hutch I drove past the retirement home where I believed Mrs. Graber lived. I stopped in and looked at the names on the board and decided to stop by her apartment.

She answered and was very gracious in answering my questions. She said that the stone was now at her daughter, Helen's place, also in Hutchinson. I got her number and gave her a call. She invited us over to

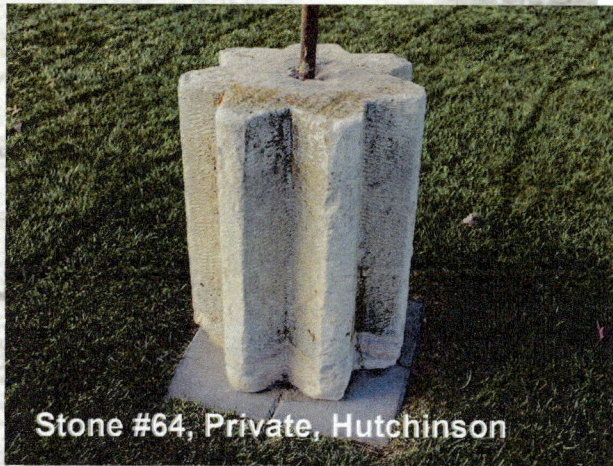

Stone #64, Private, Hutchinson

photograph and talk about their stone.

The stone was displayed upright in the back yard setting on a concrete pad. The stone had been from the Arbuckle family farm between Hutchinson and Nickerson, then moved to Pretty Prairie, Kansas, in the 40s where it was well known, until moving to the retirement home in Hutchinson and then finally to its current location.

Threshing Stone #65

The lead for this stone came from a young friend of mine. Tim Martens and confirmed later by his dad Orlin. He recalled seeing a threshing stone on a farm just east of Inman, Kansas. I often go to Inman to visit my mother and stopped at this farm on my

and unloading is heavy work that is done only by strong men. The work of moving the grain about on the floor and after the threshing, the shaking out and clearing away of the threshed straw can be done by women.

"For threshing they used a large stone, a good one meter long and a diameter of .75 meters, then with a long hole through which an iron shaft was pulled. On each end strong iron bearings were attached. The stone, called 'Ausfahrtstein', did its work by means of chiseled out 'Zakken' (ridges). The horse was hitched to this contraption and the horse had a saddle.

"A man got onto the horse and pulled this machine at a good pace around in a circle, sometimes for over an hour. The grain had to be completely threshed from the straw before I could quit. In the hot summer this was hard work for both horse and people. The straw was removed from the floor with pitchforks that had long wooden tines; then the grain and chaff was pushed onto a pile in the center of the floor. At this point another load of unthreshed grain was brought in and the process was repeated.

"All grain was threshed in this way. When the pile in the center got too large they brought in the fanning mill which, with both air flow and sieves, separated the chaff from the grain. The clean grain was put into sacks that were carried to an upstairs room. Rye was treated separately; it was threshed with a hand flail and

the straw was kept to be used as roofing material. The rest of the straw was put into stacks that needed to be carefully shaped so that rain water would be shed and the straw would not spoil. After the grain was harvested one could go on to a new task."

REV. JOHN K. SIEMENS (1892-1976)

Threshing Stones, was written by Wesley J. Prieb and printed in a brochure, unknown publisher.

"In an attempt to reconstruct the process of threshing wheat with a threshing stone, I visited eighty-one year old Rev. John K. Siemens, retired in Hillsboro,

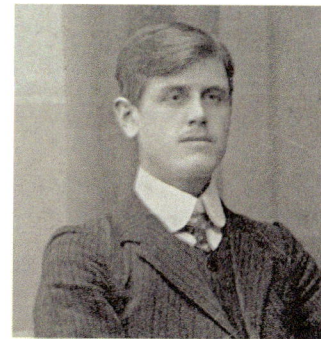

Rev. John K. Siemens, 1914 in Crimea. Center for Mennonite Bretheren Studies, Tabor College

Kansas, in the summer of 1974. Siemens came to America at the age of thirty and remembers how he helped both his father and his neighbors thresh wheat with the threshing stone. The following account is based on his experience in harvesting wheat in the Crimean village of Borangar in 1903-1905.

"Preparing the threshing floor began in spring. The floor was carefully located on level ground behind the farm buildings. A circle with a diameter of fifty feet was marked with a rope fastened to

the center with a peg. The area was then leveled and covered with chaff. The chaff was soaked with water and then pressed into the ground with a heavy roller. This process was repeated until the crust of the floor had the composition of adobe brick, but not quite as hard. While not in use the floor was protected by a layer of straw to keep it from cracking.

"The wheat in the field was cut by a reaper pulled by horses. The wheat, cut about four to six inches from the ground, was pushed by a reel onto a platform behind the sickle. A man on the platform would form a stack and then drop the stack along rows across the field. The rows of wheat were then formed into caps, which resembled wheat shocks so common to the Kansas landscape in the twenties and thirties. The wheat would remain in these caps till the day of threshing.

"On the day threshing operations began, the caps were hauled to the threshing floor with V-shaped wagons. Siemens said two horses sometimes could hardly pull the heavy load across the fields. After the threshing floor had been swept, the wheat straw was scattered on the floor to a depth of about sixteen inches. The entire floor was covered except a six-foot circle in the center.

"The threshing stone, pulled by two horses, was then rolled over the straw. The driver, riding the left horse counter clockwise, started on the outer edge of the circle and moved towards the center, covering each swath two times. The first round pressed the wheat down. During the second round the wheat would begin to shed. Turners, using wooden forks made by Crimean Mountain Tatars (Turkish ethnic group), turned the straw, also moving towards the center of the circles.

"When the horses reached the center of the circle, the driver moved the team to the outside of the circle and started the second cycle. Again the stone rolled over the turned straw twice, moving toward the center of the circle. The turners also moved to the outside of the circle. During the second cycle two to four men, and sometimes women, used their forks to shake the straw. To move the straw to the outside of the circle around the edge of the threshing floor, one man pulled a rake with tines that were about eight inches long and about four inches apart. The spread of the rake was about four feet. The dual process of shaking and raking separated the straw from the wheat and the chaff.

"After the straw had been moved from the floor, a board which served as a giant hoe or grader, pulled by a horse, was used to move the wheat and chaff towards the center of the circle. Two wagon loads of straw produced about fifty bushels of grain. The grain and the chaff remained in the center of the floor while two more loads of wheat straw from the field were scattered on the floor. The threshing process began again. This process was

Stone #65, Private, Rural Inman

way there one time. The stone was in the yard but no one was home. There is a second house on the farm so I went there and no one was home. I did this two more times at both houses before I found the owner home one day.

The stone sets at the end of the driveway as it enters the yard between the two houses. It is in good condition and has no axle hole. He says the stone has always been on the farm.

Threshing Stone #66

The "Newton Kansan" ran an article about my project and after publication I received a call from a Newton resident, Harry Kasitz. He said, "Yeah, I have one of those stones!"

I called him and set up

a time to meet with him and photograph the stone. He was very welcoming, and we went around to the back side of the house to show me the stone displayed vertically on concrete. This first thing I noticed was that it was smaller than most stones. I got out my tape measure and it was smaller; this stone was only about 28" x 22" in diameter.

Stone #66, Private, Newton

He invited me in his home and we talked for a long time, he was very interesting. This stone is one of two stones that were from his wife's family, the Bestvater farm was northwest of Goessel. I also got a picture of his wife Grace and the two stones in front of the house.

repeated three times each day, once in the morning and twice in the afternoon. At the end of the day about 125-150 bushels of wheat and chaff were piled in the center of the floor. The straw was hauled away with a net - a wooden frame with ropes stretched across pulled by a horse to the top of the straw stack and then unloaded.

"The chaff and wheat were then covered with canvas in case of rain. Early next morning, while it was still too damp to thresh, the fanning mill was moved onto the threshing floor. The fan, propelled by a man or a woman, separated the chaff from the wheat. The wheat was carried in sacks or carts to the place of storage in the granary. The chaff was hauled away with the net. Siemens said the chaff pile was very well shaped to preserve it for cattle and horse feed. Straw was used mainly for bedding for livestock and fuel.

"According to Siemens this method of threshing wheat in 1903-1905 in South Russia was basically the same as that used in 1874. This was the method the Mennonite immigrants of 1874 expected to use in America, but they soon rolled their stones into gullies to stop erosion and turned to better methods of threshing wheat."

ANNA DOERKSEN BARKMAN KORNELSON (1854-1937)

Following is an article "The Journal of Anna Doerksen Barkman Kornelsen (1854-1937): Woman of Strength," with introduction and annotations by grandson Ben B. Dueck, published by *Preservings* Special Double Issue - Part Two No. 10, June, 1997-the Newsletter of the Hanover Steinbach Historical Society Inc. She immigrated to Canada with her parents in 1875. (Note: This is not the Anna Barkman that emigrated to Kansas.)

"Economic Changes: In 1866 all landless families (*Anwohner*), including my parents were given 12 desjatien of land each (approximately 32 acres), and now I had to help my father with the plowing. Since this land was ten verst (1.067 kilometers) away from our village, we would stay on the steppe-land for the whole week, day and night. This was repeated during harvest-time in late summer.

"At that time, there were no reapers or binders; the grain was cut with a scythe and bound into sheaves by hand.

"Then the sheaves of grain were laboriously hauled home. We left home at two in the morning, or even earlier, and came back with the first load of

Anna Doerksen Barkman Kornelson. Global Anabaptist Mennonite Encyclopedia

grain by seven. We fed the horses and had breakfast, unloaded our rack and were off to the steppe for our second load. With this we returned at past noon. After lunch, we emptied the rack and went out for our third trip. We came home late; before all the chores were done, it was usually eleven o'clock. Thus, with the hauling of three loads we ended our day.

"After a few hours of rest, we were up again. This was very hard for me. Driving to the steppe, I lay down on the wagon and slept as well as I could. Going home with the full load, father slept while I had to take the reins. During the second and third trips, I took along some knitting. No time was wasted.

"This was how I spent my youthful summers. Those farmers who did not have very good horses could only haul two wagon-loads per day, and this is what most of them did. However, when the weather was dry we always made three trips daily. In our journey we had to pass through Wolf's Creek. It had very steep banks, preventing us from loading our racks very full. But, thank God, all that is behind us.

"Threshing: When all the grain was finally home, we threshed it. Horses were hitched to a specially-made threshing stone and the sheaves were threshed with this. After threshing the grain, we shook the straw with rakes to separate the kernels from the straw; this was called 'raking out.' The straw was then carefully piled while the grain with the chaff was pushed into the barn, where the grain was cleaned of chaff by running it through the fanning mill two or three times. The clean grain was then carried up into the loft."

MARGARETHA (WOELK) FRIESEN (1871-1965)

This memoir was given to me by Fern Bartel (Margaret was her grandmother's sister). Friesen's memoir was compiled by Elsie Duerksen Schmidt in about 1980 titled *Life Story of Our Parents*. Her father was Abraham Woelk

Margaret (Mrs. B. A.) Friesen, nee Woelk - circa 1902. Photo courtesy Fern Bartel

born in Rudnerweide, in South Russia on September 29, 1840. On October 27, 1863 he married Justina Friesen. Then in 1873 they moved to Sagradofka, to the village of Alexanderfeld. They wanted to move to America because of the draft. The family eventually arrived in Kansas in 1893 and settled in Marion County.

"The seed was still broadcast by hand. The seed was put into a bed sheet

Two threshing stones with Grace (Bestvater) Kasitz. Photo courtesy Harry Kasitz

The Bestvater house is now located on the Goessel Mennonite Heritage and Agricultural Museum grounds.

He said when he goes he wants this stone to be given to his nephews for display at the Old Mill in Newton.

We have not located the other stone yet, but we are still looking.

Threshing Stone #67

This stone had several leads that all ended up being the same stone in response to an article in Thresher E-View.

I received an email from Ann (Schmidt) Pasnick saying "Interesting project! We have a threshing stone on the family farm north of Pawnee Rock, Kansas. Dad got it from his sister Irma Schmidt in the 1950s when she and her husband John moved from their farm north of Newton to Washington state.

"Just a month later I was contacted by Stan Schmidt with this information, "My dad, John Schmidt was the youngest of 10 children of P. A. Schmidt, who homesteaded the farm. My dad bought the farm from his dad and farmed it until 1957 when he had a farm sale and moved to Washington. I remember clearly that we had a 'perfect' threshing stone at our farm until sale day.

Stone #67, Private, Rural Pawnee Rock

which in turn was thrown over the left shoulder and the seed was broadcast with the right hand. Father had a difficult time doing this because of his sore leg. Often compassionate neighbors would help out.

"Harvesting was done with the scythe, with Russians usually doing the cutting. Also strong Mennonite boys also knew how to handle the scythe. The cut grain was raked together by women into rows and straw was used to tie the grain into bundles. The young children considered it a high point if they could go along when faspa [the Low-German name for an afternoon light meal] was taken to the field where they could drag the bundles together where father would stack them.

"The threshing was done with threshing stones. The bundles were taken home by hay racks (called *leiterwagen).* The bundles were taken to the threshing floor where the grain was then spread even on the threshing floor. Two horses were hitched to the threshing stone and with two threshing stones they went round and round. They (straw) grain had to be turned several times until all the grain kernels were knocked out.

"Small children were used to ride the horses round and round. One would sympathize with the small children when they were routed out of bed early in the morning and when they were scolded when they got sleepy in the mid-day heat on the horse they were riding. The empty straw was raked by practiced rakers into tight rows of which my mother was a pro-

fessional. A pointed stick called a Bacurr was used to spear the row of straw, then one of us girls would help to upend the stick and straw, which then was carried by the Russian boy up the straw stack.

"Often times we also had to help. We did this in horse blankets. This was terribly hard work. The straw stacks were all put up in the yard. If the Russian boys did a good job the stacks looked very neat. The wheat straw was used for heating. The barley straw and chaff were used for cattle fodder.

"The threshing took weeks to complete. After this the grain had to be cleaned by hand with a fanning mill which usually took two weeks. It had to be fanned three times. The chaff was stored upstairs in the barn. The cleaned wheat was stored upstairs in the house."

Heinrich A. Woelk (1874-1960)
The following is a letter from Mrs, Margaret Friesen, an account of farming in Russia as told by her brother Heinrich A. Woelk written in 1936 in the German language and translated by J. A. Duerksen. [This account gives great insight into farming practices of the time.]

"Now I want to tell a little about farming in Russia. Usually we had our first snow in November. The horses and cows were kept in the barn during the winter months. We had room for 13 horses and eight cows. We got up at 6 o'clock in the morning. First I went in the barn to wake our Russian servant. Then we started our chores. The first chore was to water the horses. The well was in the barn.

"One of us drew the water out of the well and poured the water into a large barrel while the other one watered the horses. We had two large cribs, they were water tight and were about 12 feet long and 4 feet wide. Four horses were tied on each side of the crib. We poured the water into the cribs with a bucket. This chore took quite a long time. We fed the horses and the cows. Then the Russian servant carted the manure out of the barn to a manure pile in the yard. This large manure pile remained on the yard until spring.

"Once a week we cut barley straw with a machine possibly similar to an ensilage cutter. This took half a day. This lasted us for about a week. This routine is followed during the whole winter. The harness for the horses were repaired and oiled before seed-time.

Heinrich A. Woelk.
Photo courtesy Fern Bartel

"Then the wheat and barley were fanned so that the seed would be clean for seeding. After all these preparations we were ready for sowing. We usually started sowing in mid-March. We had two plows, *schratzpflug*, five horses were hitched to this plow. The two horses in the rear were hitched directly to the plow; a chain ran from the plow to the front between the two horses in the rear; the other three horses were hitched to this chain. The driver sat on the last rear horse. He had reins to the front horses with which he guided them.

"Seeding time lasted about two weeks. A citizen's meeting was held a short time before seed-time. At this meeting they discussed and finally agreed on which fields they would sow wheat and on which barley and on which land they would plant corn and which fields they should lie in fallow for the next year.

"After seed-time was finished we started to haul the manure away, first all the dry manure was peeled off the large manure pile. This dry manure was hauled to and spread over the fallow ground. The wet manure was cut into bricks with the spade. These bricks were placed on the flat ground to dry. A few days later the bricks were turned around to dry on the other side. After a few more days the bricks were put into towers up to five feet high so that the wind could blow through the bricks and dry them out completely. After the bricks were dry they were carried into the bin next to the furnace. Then the bricks were ready for heating the furnace next winter. After the bricks were finished the fallow land was plowed.

"We had a one-share plow. The tree was supported by the wheels. There were two handles at the rear end of the plow. One man had to walk behind the plow guiding these handles. Seven horses

"After that I lost track of it. Nobody in their right mind would think of moving one of those heavy stones, so it may be there at this moment. If so, it would be south of the house, and south of the outside kitchen under the cottonwood trees. Over the years I remember seeing the wood framework rot and gradually fall off the stone, but it was a complete working tool in its earlier day. My dad sold the farm to Ferd Schmidt in 1957."

Now, however, we still have a mystery. Stan remembers a wooden towing frame, yet the three photographs sent to me by Ann, show no axle hole in the stone. Curious!

Threshing Stone #68

Stone #68, Private, Rural Goessel

I first heard of this stone from Brian Stucky at Goessel Threshing Days that Rudy Voth had a threshing stone on the farm 2 miles south of Goessel. I contacted Tim Voth, his son, who lives on the farm now to make arrangements to photograph it. It has always been on this farmstead where it sat on its side when photographed on a cold winter day.

Threshing Stones #69 and #70

Stone #69, Private, Newton

I found these stones by asking the right person. I share an interest in local Alta Township history and Alta Mill history with Jerry Wall who grew up on Dutch Avenue just a mile

were hitched to this plow. We plowed quite deep. I think up to eight inches deep.

"The corn was planted next. Harvest started in July. We had a *Haspel* machine. There were two small wheels in the front part of the machine to which the pole is attached. We hitched two horses to this machine. The seat was above the two wheels and the driver who guided the horses sat on the seat. The machine had a big platform. A man with a fork sat on a seat at the rear of the platform, this fork had three teeth. One of the teeth was quite long, the second tooth was shorter and the third was quite short. The Haspel threw the wheat or barley on the platform. As soon as there was quite a pile of straw on the platform he shoved it backward off the platform on the ground. There were usually two men following the machine straightening the piles.

"Immediately following cutting the grain we started threshing with the threshing stone, *utfoasteen*. We had two threshing floors. One threshing floor was for threshing wheat the other for barley. The threshing floor was a circular level piece of hard-packed ground. The threshing floor was put in order before harvest set in. Then the ladder wagons were put in good running order. Two men would go in a ladder wagon to the field to haul a load of grain to the yard at home. One person was on the wagon loading the grain while the other one pitched the grain with the fork onto the wagon. Only the man on the ground went back home with the

load of wheat or barley, while the one on the wagon remained in the field. He had to rake the grain with a large rake. He had to be ready by the time the ladder wagon came back again. The rake was about six feet wide. It was always very interesting loading the wagon.

"We threshed about eight loads a day. We had two threshing stones. Those who worked at home on the yard had to have the threshing floor clean by the time the wagon came. The straw was raked in tight rows. We had a long pole. It was called a straw-pole. One end of the pole had a long belt fastened. The pole was stuck into the raked straw with the pointed end, the other end with the belt remained clear. When the stick was full it was raised up. He placed the bottom end on his shoulder and carried his load to the straw stack. After the straw was cleared off the threshing floor the grain with the chaff was shoved into a bin with a four-foot sledge. The threshing floor was clear before the next load came home. It took several weeks to thresh the grain.

"Next the grain was fanned in a fanning machine, separating the grain from the chaff. One person turned the fanning mill while the other person shoveled the mixture of grain and chaff into the fanning mill with his wooden fork. One person sat in front of the fanning mill to shovel the grain away from under the fanning mill. The Russian servant carried the chaff in a bed-sheet to the second floor of the barn. After the second fanning of grain the grain was carried in sacks onto the second floor of the house. That was hard work.

"When we were ready to sell the grain a Jew would come around to our house to buy the grain. He brought his own sacks. We filled the sacks and then weighed the filled sacks on a large scale. These sacks were then sewed shut and placed outside on the yard. There were five Pud (about 180 lbs.) in each sack. Forty Russian pounds was one pud.

"Then in fall we plowed as much land as possible. Then came the time for butchering hogs. That was always a real feast for the children; we were often allowed to stay home on butchering day. We children always were ready to run for things that were needed. We usually butchered four or five hogs. For a week or more our parents helped their neighbors butchering hogs.

This is all I have to say."

AUTHOR UNKNOWN

From North Dakota State University Libraries-Germans from Russia Heritage Collection, *Bessarabia: The Homeland in Pictures*, Pictorial Calendar 1987 -Translation by Alma M. Herman June 1991.

"Farming was hard work in July, but it also had its pleasant side...the cut grain had to be stacked in *kopitzen* (heaps) with wooden forks, as always. The grain was carefully gathered with large rakes.

"Threshing time followed the harvest. The cut grain was brought to the threshing place with the *harbi-wagen* (large harvest wagons).

"In the farmyard, as a rule, was a well from which water, that precious liquid, was drawn up with a pail and a reel. People and animals were refreshed at the well when threshing made them thirsty.

"Life on the farm was ruled by regularity during the warm months of the year. Man and beast were tied into the daily routine. Household work was gladly performed outdoors, and children played under the watchful eye of the mother.

"At the beginning of June, grain harvesting began. School vacations had begun. Young and old, including the mother, were on the steppe from early morning to late evening.

"Four to six *hektar* (12 acres) were harvested per day. The cut grain was set up in bundles and raked together with large wooden rakes tied with rope of twisted straw, then stacked into kopitzen held together with flat rope.

"Tired? No, yet everyone was. Only the young Russian girls sang their melancholy songs on the way home, their voices ringing out far over the steppe.

"And then came the threshing time that lasted three or four weeks. Early in the morning, sometimes at 2 o'clock, the rumbling wheels of the harbi-wagen began to roll by, the noise lasted only half an hour, then, sleep returned. Around 8:30 in the morning, the highly-loaded harvest wagons drove leisurely into the yard. I can

from my home place. I knew he was a person to ask if he knew where any threshing stones might be located.

I ran into him at a local restaurant and asked him if he knew of any and he responded with, "My mom has two at her place here in Newton." He said both came from his parents farm, Ray and Ruth Wall, where they sat in the cattle corral used as salt block holders.

Stone #70, Private, Newton

They both were part of the Wall side of the family and her mom moved one to her place from the farm when she moved to Newton.

The other stone was moved to Burton, Kansas, for a while, before moving to its current location in Newton.

Both these stones have the tops hollowed out, typical of those used for salt lick stands and show extreme salt deterioration from many years of use. Both stones were setting upright and covered in snow the day I photographed them.

Threshing Stone #71:

Stone #71, Private, Hesston

I learned of this stone directly from the current owner. This stone was originally from a farm along Dutch Avenue very near to where I grew up. It has been modified for use as a salt lick stand and has no axle hole.

still hear my brother Jakob's loud 'Brr' at the sight of two or three highly stacked harvest wagons in the yard. The horses were unhitched and fed and only then was it breakfast time.

"Meanwhile, the sun over the threshing place was high enough in the sky so that the grain could be spread out on the threshing floor and threshed by dragging over it three heavy stones, each drawn by two horses.

"Who among us doesn't know what it meant: to lay-out; turn-over; rake; carry straw; pitch up high; set a straw stack; gather up the *drusch* (threshed grain) with a sled; to run it through a cleaning mill; clear space; gather chaff; measure grain and transport it on *bühne* (platform) to the *dachboden* (loft). All that comprised a threshing day.

"As a child, I liked best to clear away the flaxseed. It was so cool, smooth, and soothing; its smell was superb and one could 'plop' into it so nicely. Yes, one could gain something fine from every kind of hard work.

"After a day of such work one did not need a sleeping pill to fall asleep. Big brother and the hired man set up their night sleeping quarters out in the open in a lay down *grogwagen* (trough wagon), in order to be at their post promptly in the morning. Before their departure, as mentioned, at 2 o'clock in the morning there was no breakfast, but *nussaschnaps* (a sort of whiskey) was not frowned on because it helped one stay awake during the long ride.

GEORGE BECKER (1911-2012)

Meeting George was the most surprising discovery in all my research on threshing stones. As I stated earlier I should have started this project a hundred years earlier, and little did I ever expect to personally interview a person over 100 years old, a person who actually used a threshing stone in the Ukraine as a young boy.

My research uncovered the name of a man in Inman, Kansas, who might have owned a threshing stone, George Becker. I heard he lived in the Pleasant View Retirement Home in Inman where my mother lives.

George Becker.

One day I asked mom if she knew him. She said, "Yes he lives in the building and he was a good friend of dad." I walked over to his apartment to see if he was in. Soon an elderly man answered the door and he quickly invited me in to have a seat. I introduced myself and who my dad was, and he had great things to say about my dad.

The first amazing thing was how much he reminded me of my dad, in looks, sound and his quick wit.

I soon got to my question of whether he had owned a threshing stone and he said, "No, unfortunately not." We chatted for a few more minutes and I got up to leave. As I approached the door, he said, "But we used one when I was a boy in Russia!"

What? I turned around and we talked for another hour about his memories of that experience. His memory and knowledge was amazing, and he had an answer for my every question.

He proceeded to tell me about his recollections of his young life. He was born in Russia in 1911. The family owned a flour mill, and they lived in Franzthal in the south-east corner of the Molotschna Colony.

[But let me first set the context of his recollections. Many Mennonites in the Ukraine were nervous about their situation in the country at the time George was about 6 years old. Many Mennonites greeted the resignation of the Tsar of Russia, Nicholas II in February 1917 with consent and considerable relief. The tremendous anti-German pressures of the war years had taken their toll even among former patriots. Others realized, to be sure, that violent revolution, if it came, would exact heavy a heavy price of all people in Russia. Mennonites would be no more exempt than any others.

In only a few months Halbstadt and other villages of the southern Ukrainian Mennonite settlements came to feel the full impact of terror and bloodshed that would soon engulf the entire country. Outlying estates were attacked and burned while their owners fled to the safety of the main Mennonite colonies nearby. The final defeat of the White armies in the fall of 1920 terminated all effective opposition to the revolution. Many Russian Mennonites made their way eventually to the United States.[1]

The Becker family, along with other families were not immune to these troubles either. The military had taken much of the farming implements to use for metal in the military effort. So the threshing machine was not available for use to thresh the wheat harvest. Many threshing stones were still laying around in the fields that had not been used for years. But many older farmers still remembered well how to use them. They got them out and fixed them up to be used once again.

George was one of the young boys that rode atop one of the two work horses as they towed the threshing stone(s) over the grain and straw. He recalled the threshing circle to be about 50 to 70 feet in diameter. The threshing floor was made of dirt, smoothed and then packed and watered. It was so hard that the threshing stone would not make a dent in the surface.

The grain was actually cut a bit green and then threshing followed later at the farm yard. The grain was piled to maybe a foot or less deep. The horses

Threshing Stone #72

I was contacted by Walter Lehman after he read about my project. He said there was a stone at his home place where his parents still live. I contacted Brent and Karen Lehman and made arrangements to photograph the stone setting upright on the back patio. Galen Dreier said the stone was his grandfather's Albert Steiner and has always been on this farmstead.

Stone #72, Private, Newton

Threshing Stone #73

I had driven past this farm many times in the Inman community and never saw the threshing stone located in the front yard, not obvious from the road but could have been

seen. As I have been driving around the local areas over the last year I am always scanning for threshing stones that I may spot on a yard, but I missed this one.

Stone #73, Private, Inman

I got an email from David Balzer, the current owner of the farmstead on which this stone has always belonged. It is from his great-great-grandfather Peter Balzer who first farmed the property. It has always set horizontally on the lawn in the front yard for as long as anyone can remember.

Threshing Stone #74

Phil Considine contacted me after reading an article about my project in a local

worked at a walking pace, as they went round and round from the outside of the circle to the center, over and over. As the two horses pulled the stone, a bunch of men with wooden forks turned the straw to the outside of the circle. The straw was piled and used for heating and feed for cows and horses, while the chaff made good feed

He recalled it was very hot, and the adults watched out for everyone to be sure no one collapsed in the heat.

The threshing stone was towed with a wooden frame. He recalled that the threshing stones were made on the outskirts of town, and they had to be perfect. He remembered the reason for seven teeth was that it was a biblical number.

They used horses to pull the stone, to his recollection they were Belgian horses. But they also used Bortrin, but he was not sure if he knew the difference. He added that the village had one stud horse and one bull for the whole community to use, things back then were very communal. The horses they used for "driving" [carriage pulling] were a different and smaller breed. They also used oxen for farm work. The oxen were cheaper to buy and feeding was cheaper too. He recalled that at one time they even used some camels on the farm when enough horses were not available.

The horse was controlled with a small whip, and controlled with words like "gee" and "haw" for left and right and "pdurr" for stop and "noporre" to go;

oxen were controlled by "sub and subea."

Sometimes the horses had to go to the bathroom while they were threshing wheat, so they had to take the fork with some straw and scoop it to the side. The saddle he rode on was made of rope and some gunny sack.

Eventually the wheat was winnowed, and the gunny sacks of wheat went up into the attic of the house to help insulate against the cold of winter.

In 1920 his father sold the flour mill, and his family started their migration to the United States. The trek took two years, spending about one year in Crimea via Yalta near the Turkish border. When he was 11 years old they arrived in the U.S. His father died on the trip to the U.S. and his mother could not raise him and gave him up for adoption, so he was raised and lived in the Inman community.

I had a few follow-up questions to ask George, so on July 4, 2012 I stopped in to see George again, but he had passed away, just the week before, shortly before his 101st birthday.

Turkey Red Wheat

The story that the Mennonites brought Turkey Red Wheat to Kansas and transformed the prairies into the Breadbasket of the World is well known in Kansas history. However, the real story is not that simple. There was hard winter wheat in Kansas before the Mennonites came, the Mennonites didn't bring enough wheat in their trunks to make this transformation, and hard winter wheat was initially not popular because the flour mills couldn't mill it. These issues have long been heatedly discussed by historians and agronomists. It is fascinating to hear the various and numerous arguments that have been published discussing this issue over the last one hundred and thirty five years.

"Earliest reports of wheat as a crop in the area now known as Kansas are found in records of various Indian mission farms and reservations in pre-territorial days. The Shawnee Methodist Mission reported sowing 100 acres of winter wheat in 1839. The Sac and Fox farmers sowed 40 acres in 1850 and Indians of the Osage tribe were reportedly sowing wheat in 1851."[1]

The kind of wheat to plant was actually a subject of much debate in Kansas prior to the Russian-German arrival. Most early settlers preferred corn for its greater household use and as feed for livestock, especially pigs. Land promoter T. C. Henry was one of the first to plant winter wheat on a large scale, in virtual plantation style, near Abilene in 1873. The question of which would be the dominant grain for Kansas was actually being settled upon the Russian-German arrival, and the grasshoppers deserve some of the credit—for wiping out the corn crop and most of the spring wheat. Only winter wheat was generally successful in 1874.[2]

When farmers first came to the plains of Kansas in the mid 1800s, they planted their wheat in the spring. Here they were following the practices of their ancestors, who had farmed in the Ohio Valley, the mid-Atlantic states, and before that, in Europe. As these pioneers moved farther west into Kansas, their crops began to suffer and their yield declined.[3]

Turkey Red Wheat is a cereal grain that starts its genetic life in the strains of wild grasses growing in the Fertile Crescent region. It evolved into what would later be called Turkey Red Wheat. This wheat type was experimented with by the Mennonite farmers particularly in the Ukraine.

The Mennonites called the wheat Turkey because it first grew in a little valley in Turkey where they first obtained it.[4]

Opposite - Kansas became the "Breadbasket of the World" - Kansas wheat field in southeastern McPherson County - 2012.

It came with many local names, including, for example, Crimean, Malakof, Red Russian, Red Winter, Tauranian, Turkey Red, and Kharkof. All of these (and twenty-one other names) were considered to be synonyms of Turkey.[5] Although all are similar morphologically, no doubt there are differences among them due, in part, to the places of origin. In other words, all of them did not come out of the same sack. Therefore, we should consider Turkey as a type rather than a specific variety with narrowly definable characteristics.[6]

Turkey is a winter wheat, considered to be mid-season for maturity – late when compared with the varieties being grown now. The stems are white at maturity, slender, and weak, with a tendency to lodge or lay over if growth is heavy. The leaves are narrow and dark green, and the plants are winter hardy and drought enduring. The heads are bearded and have predominantly white chaff. The grain is dark red in color and hard in texture. The variety is not resistant to common diseases, but some selections have some resistance to bunt or stinking smut, and to rust. The variety is susceptible to Hessian fly damage.[7]

The Russian-Germans were also accustomed to planting spring soft wheat in Russia, in the case of the Molotschna

Turkey Red Wheat Grain - Grown by Virgil Litke in 1974.

Colony, a soft wheat called *Girka*.[8]

Only very small quantities of a "spring hard" red wheat were planted in the Ukraine. The *Arnaulka* or *White Turkey* was grown in the Volga region as a spring wheat, and was probably the kind that was adapted for winter planting in Kansas within a few years after the Russian-German arrival.[9]

Turkey Red's reputation was for being a hardy variety that could withstand the harsh cold and resisted drought.

Years of use of this grain in the Ukraine helped to develop strong demand and export markets. Farmers were no longer just raising wheat for their own consumption but growing large quantities for export to European markets where the grain was very desirable. One source reported that they had been shipping 10,000 bushels annually to London and Liverpool and that "there wheat brings ten cents per bushel more than almost any other."[10]

As the farms evolved from generation to generation, the acreage for each farm diminished. This combined with losses of religious freedoms inspired the Mennonites to explore other farming opportunities around the world.

With the strong promotions of the State of Kansas and the Atchison, Topeka

paper. The stone was from his wifes Kathy's family, it had been on the property of her great-great grandfather Carson Reed of rural Burrton, Kansas. The stone is displayed in a flower bed near the road, it has been re-purposed with a horse tie ring.

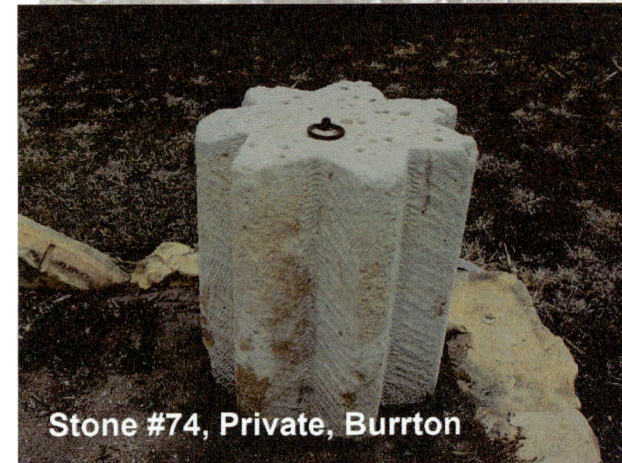

Stone #74, Private, Burrton

Threshing Stones # 75 and 76

These stones were first pointed out to me by a friend of mine, who said he knew a person who may have several stones. Additionally, I have been told by numerous individuals that this collector had many

Stone #75, Private, Moundridge

stones, one man even thought that he had seen up to 20. I tried several times to contact the family but never got a call back.

I could easily see the two stones located on the front porch by driving by the residence. But I certainly did not want to photograph without permission.

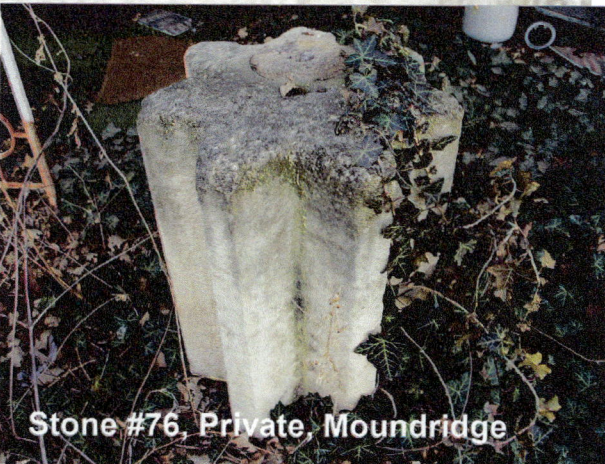

Stone #76, Private, Moundridge

and Santa Fe railroads, many looked to the Kansas plains as the ideal place to relocate. The climate, rainfall and soil were very similar to what they were familiar with and the Turkey Red wheat appeared to have promise in the new country.

The great migration of 1874 started with families settling in the Gnadenau settlement just east of current day Hillsboro, Kansas. Credit is given to German Mennonites who settled here in August of 1874 for being the first to bring the Turkey Red wheat to Kansas.

The most complete accounts of the Mennonite migrations and their association with Turkey Red wheat are the reports of David V. Wiebe, Cornelius Krahn, and James C. Malin. In *They Seek a Country: A Survey of Mennonite Migrations* all three gave abundant evidence that insofar as the Kansas Mennonites are involved, Turkey seed wheat was introduced and first sown in 1874 in Harvey and Marion counties. By the end of 1874, Wiebe states, 600 families went to Kansas, 80 to Nebraska, 200 to Dakota, 15 to Minnesota, 230 to Manitoba, and 150 remained in the East.[11]

The Mennonites were able to succeed as farmers on the plains while many other American colonists failed. Mennonite advance agents had sought land and climate similar to that of their home country in Russia, and they found these in Kansas and Nebraska. They had experience in farming on prairie land and, therefore, were not dismayed by what

Kansas wheat grown by Wayne Schrag - 2012.

they found. They brought along some suitable equipment, know-how, and, in the case of wheat, seed of a type that was well-adapted to the environment. No single family or small group of families should be given exclusive credit for the introduction of Turkey wheat. Surely there is nothing wrong with passing deserved credit around.[12]

The first party who came to Kansas brought, all told, not more than twenty or thirty bushels of seed wheat of the Russian or "Turkey" variety, which had been the most popular with them in their Russian home.[13]

When this strain was planted in combination with improved tillage practices, such as dry farming, fertilizing, and deep plowing, the farmers of Kansas prospered, and Turkey Red wheat became a grandparent of all Kansas wheat today.[14]

Here is where the myth and fact start to become blurred. In the spring of 1874, Anna Barkman, then an eight-year-old girl living in Caslov, Russia, hand-picked 250,000 kernels (2 gallons) of Turkey Red wheat. This seed was brought to Marion County, Kansas, by the Barkmans, sown in the fall of 1874, and harvested in 1875.[15]

In 1927, historian Bliss Isely, an experienced Wichita newspaperman, asked two Bethel College people, a professor, David Richert, and a student, Herb Schmidt, to become involved in the subject of Turkey Red winter wheat as research for his booklet, *Early Days in Kansas.*

They drove to Hoffnungstal village near Hillsboro to interview Anna Barkman Wohlgemuth. As recounted in a 1974 interview with Herb Schmidt, by Ray Wiebe, Schmidt recalled that they had an enjoyable talk, alternating in High and Low German language. "She had helped prepare for the trip to America by picking individual kernels of wheat and placing them in a cloth sack. The sack was placed in an old-fashioned box and packed with other personal effects in her parents' trunk... adding that other families had followed a similar procedure at the time."[16]

Turkey Red Wheat. Virgil Likte.

This account was then printed in the "Bethel College Collegian" and also the May 15, 1927 issue of the "Wichita Beacon." From these publications the story was launched.

The story has been told in many ways over the years. It was included in the Isley and Richardson history books for Kansas school children from the 1920s to the 1960s. It was a dramatic reading on a radio show produced in 1941 over the Kansas State Network of the Mutual Broadcasting System. It has been made into a coloring book for children, presented in many historical accounts, and even made into several theatrical productions in 1974.

Following are excerpts from the 1941 radio drama.

-Anna – Fifty-three --- fifty-four. Nope, color's not right --- you can't go with us to America. (pause) Fifty-five.
-Mother – And how are you getting along, Anna?
-Anna – All right, Mother. But it's pretty slow work.
-Mother – I know, dear, but your father wants only the best seed to go along with us.
-Anna – Yes, Mother, I know. And I'm being very careful, too.
-Mother – You remember all your father said?
-Anna – Oh, yes, indeed. We must take only those seeds which are the right color of red and are hard, like flint.
-Mother – And you must be careful not to get any which are malformed.
-Anna – Oh, I know --- they must look just right.
-Mother – That's right, Anna.
-Anna – (Pause) Mother, when do we start

I finally got a hold of him, explained my project and asked for permission to access the property and photograph the stones that I found. I got permission. He said his father had collected them. I asked if he knew of others on the property or in his family possession and he did not.

I photographed the two on the front porch, setting upright and covered in ivy. Then I then went around to the back of the house and found 3 more covered in foliage, after first walking right past them. These are stones 80, 81 and 82.

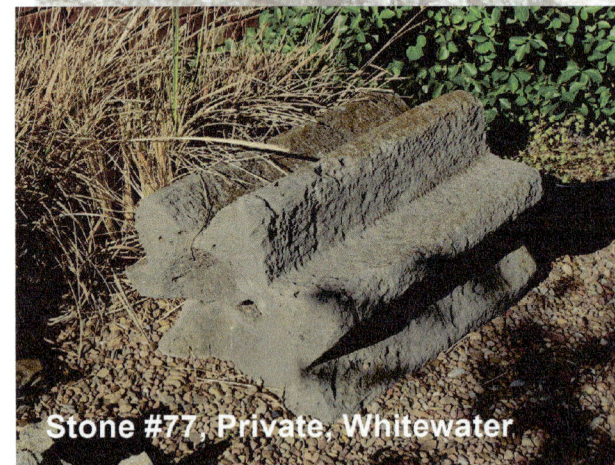

Stone #77, Private, Whitewater

Threshing Stone #77:
I was called by the own-

er of this stone now living in Whitewater, Kansas. It is from his family farm near Buhler, Kansas. He has this stone displayed in front of the house on its side. It has been hollowed out for use as a salt lick stand, and has a crack through the middle of the stone.

Stone #78, Private, Sedgwick

Threshing Stone #78:

This stone came to my attention from a friend of the currrent owner after he read about the project in the paper. I made arrangements to photograph the stone on a farm northwest of Sedgwick.

The owner believes the stone was on the farm when his family bought it, about 100

for America?

-Mother – In just a few days now.

-Anna – Why are we going?

-Mother – Because the Czar wants us to become Russians.

-Anna – And we don't want to, do we, Mother?

-Mother – No, Anna, we don't.

-Anna – I think I'm going to like America.

-Mother – You'd better get on with your seed work or we'll never get started.

-Anna – Yes, Mother. (There is a plink of seed in the can) Fifty-six.

-Mother – And remember Anna, you'll get your reward at the end of the week --- one whole handful of hazelnuts.

-Anna – Yes, Mother, thank you. (another plink) Fifty-seven.

-Narrator– Thus the Mennonite children carefully s elected each seed --- one at a time --- until they had hand-picked 259,862 grains. That way the Mennonites assured themselves that none but the very best of their Turkey Red Wheat would be brought to Kansas. Mr. Schmidt of the Santa Fe also arranged their rail transportation to their new Kansas homes without charge. Peter Barkman, his wife, and Anna arrived at their new home in Hillsboro, Kansas. One night, Barkman was talking with a neighbor who had been a wheat-grower for many years.

-Farmer – No sir, Barkman --- I think you were plumb crazy to come to this state.

-Barkman – I do not think so.

Anna Barkman family trunk that carried the Turkey Red wheat to Kansas - Donated to Adobe House Museum by Dietrich G. Barkman Family. Center for Mennonite Brethren Studies, Tabor College

Here we have freedom to think as we like, worship as we please.

-Farmer – Yeah, but you also got grasshoppers and drought.

-Barkman – We are used to dry weather.

-Farmer – But them grasshoppers --- what can you do about them when they come in great clouds like this last summer and eat up everything?

-Barkman – Everything?

-Farmer – Everything that was green and a lot that wasn't. They didn't leave enough wheat in my fields to stuff a shotgun.

-Barkman – That is too bad.

-Farmer – Too bad --- it's terrific! I don't like to discourage you new fellows ---- but

there are a lot of hardships to farming in Kansas.

-Barkman – We are used to hardships. But we always win out somehow.

-Farmer – Seth Hones, down the lane, was telling me you fellows brought along a new kind of wheat.

-Barkman – Well, yes, we did.

-Farmer – Any good?

-Barkman – We think so.

-Farmer – What's it called?

Barkman – We call it Turkey Red --- Turkey for the land where it originated and red for its color.

-Farmer – I see. Well, how's it different from our wheat?

-Barkman – Yours, I believe, is a soft wheat. This is a hard winter wheat.

-Farmer - What makes it good?

-Barkman – It resists drought better and when rains do fall, it takes advantage of them.

-Farmer – Well, Barkman, I wish you luck; looks like you're gonna need it.

-Barkman - Thank you.

-Farmer – And if your wheat, that Turkey Red, is any good – well, maybe I'd like to borrow some from you.

-Barkman – We shall see what the next year brings. If there is a good crop, we... (his voice trails off)

-Narrator – Barkman and his fellow Mennonites proved to the native Kansans that Turkey Red wheat would grow and thrive in Kansas, that its high gluten content

Anna Barkman Wohlgemuth. Center for Mennonite Brethren Studies, Tabor College

made it most desirable in the markets of the world."[17]

Anna Barkman Wohlgemuth died shortly after telling this story and follow-up accounts were not verified.

However, prominent historians debate the authenticity of this story and say it must be myth. Exhaustive, detailed books and articles have been written debating the rights to claim the origins of Turkey Red wheat in Kansas. Some give credit to the Mennonites. Others say scientists, botanists, and agricultural experts were experimenting with different strains and Turkey Red wheat was becoming the wheat of choice.

The State of Kansas was involved in exploring different crops and with both academic and governmental interest in agriculture, created the nation's first Department of Agriculture in 1855.[18] Studies at the University of Kansas and prolific publications from the university attest to the importance of wheat in this state. As early as 1862, the Kansas Department of Agriculture was seeking wheat samples from Russia as well as other countries.[19]

The alternative view includes the career of scientist Mark Carleton. Carleton moved to Kansas from Ohio as a 10-year-old in 1876. He attended Kansas Agricultural College and became a botanist. He

years ago. He didn't know if it was a Mennonite family or not. The stone had developed a large crack through the middle, over the winter. This stone was displayed vertically and had a large ring attached to the top as a place to tie horse reins.

Threshing Stone #79:

The owner of this stone read about my project in a Bethel publication. She says she doesn't know the history on this stone, other than it was on the back patio, for as long as she could remember, where she grew up in Moundridge. The wood attachments were added sometime after 1970.

It was later moved to the Moundridge Memorial Home. Later she gave the stone to the

Stone #79, Private, Elk City

present owner who has it beautifully displayed on a flat rock next to a lake on their farm outside of Elk City, Kansas.

Threshing Stones #80, #81, and #82

Stone #80, Private, Moundridge

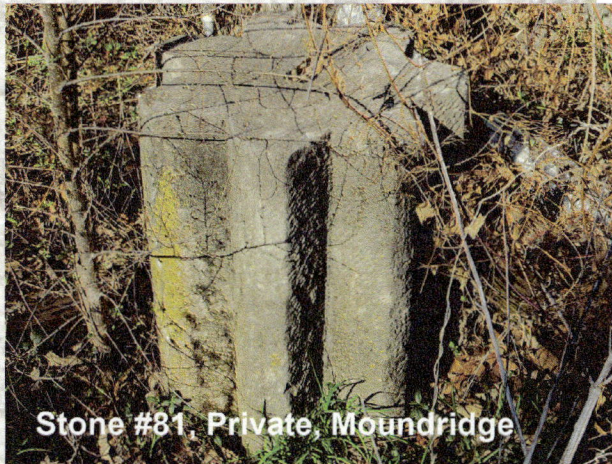
Stone #81, Private, Moundridge

taught at Fairmont College (Wichita State University today). He worked for Kansas State Experiment Station in Manhattan, and was promoted to the Department of Agriculture in Washington, D.C. He returned to Kansas to experiment with different strains of wheat to find the best variety for the hard winters and hot summers of Kansas.[20]

In a best selling book, *Hunger Fighters*, author Paul de Kruif in 1928 published the story of Carleton's ideas. Carleton saw those farmers get yields of thirty bushels to the acre while ruin raged all around them. He made friends with these Mennonites, asked them questions about the source of the wheat and was told, "Well, it was called Turkey Wheat... the first settlers, their old folks, had brought it...it's the old wheat of Russia and it does wonderful in this new land." In 1900 he traveled to Russia and returned to experiment and work with government agencies to improve the strain. In his perspective the bringing of the wheat to Kansas was inevitable; it was the scientific experimentation that made the Turkey Red wheat the dominant strain.[21]

In 1944, James Malin, a Kansas historian, took on the debate in a controversial book *Winter Wheat in the Golden Belt of Kansas: A Study in Adaptation to*

Plaque at Swiss (Volhynian) Mennonite Memorial, west of Moundridge, Kansas.

Subhumid Geographical Environment. Malin regarded the Mennonites as having some influence, but his research showed that Kansas farmers had been experimenting and questioning farming practices prior to the Mennonites' arrival in Kansas. Others, including newspaper editors, railroad land agents, land speculators, and politicians were also involved in the discussion.[22]

Furthermore, other varieties of wheat dominated the wheat harvest into the 1890s. Malin noted that "even Bernhard Warkentin, the prominent Mennonite farmer and miller, who supposedly boosted the adoption of Russian wheats by importing a large amount of seed from Russia in 1885-1886, was still planting May wheat on his farm in 1888."[23]

In the years after, each of these publications received counter-debates. However, I think Malin summed it up best in his conclusions, "Beyond the fact of bringing hard winter wheat from Russia, [the Mennonites'] positive contribution lies largely in the high quality of their farming and the shrewdness in adjusting successfully their traditional agricultural system to the new American crops, machinery and environments."[24]

The bringing of the Turkey Red wheat to Kansas, the tilling of the ground, the growing of the crop was a boon for the state, but it took about 25 years to fully materialize.

Another contributing factor to its slow acceptance was the transition taking place in the flour-milling industry during this same period of time.

The Turkey Red wheat kernel was full, plump and hard; a bushel weighed in at 60 pounds. The kernel had a reddish tint in color. The grain was harder than spring wheat, which the Kansas millers were used to. The mills, the few there were in central Kansas at that time, could not process the hard kernel.

But this did not deter the farmers and industrious entrepreneurs from taking on the challenge of figuring out new ways to mill this hard seed. Malin, in his book *Winter Wheat*, says "It is coincidence that three independent transitions were taking place simultaneously about 1870-1885. A revolution in the flour milling industry was under way...Kansas was outgrowing pioneer standards of efficiency in milling and quality of flour...and it was necessary to meet the high competitive standards both domestic and foreign."[25]

The steel roller mill technology was being developed using cylindrical rollers to crush the wheat rather than grinding them between stones such as in a traditional grain mill. The moistened grain was first passed through the series of break rollers, then sieved to separate out the fine particles that make up white flour. The middling then made multiple passes through the reduction rolls, and was again sieved after each pass to maximize extraction of white flour from the endosperm, while removing coarser bran and germ particles.

One by one the mills were equipped with steel rollers. The most prominent Mennonite miller in the central Kansas area was Bernhard Warkentin, son of miller Bernhard Warkentin, Sr., who has been credited with the introduction of Turkey Red wheat in Molotschna district in about 1860. Bernhard was influential in bringing Mennonites and their skills to Kansas; he came from Russia to Halstead in 1873. There he built and managed a grist mill for his father-in-law, C. Eisenmayer, until 1886.

In 1885, Warkentin made

Steel roller mill, pre-nineteen-hundred. Kauffman Museum

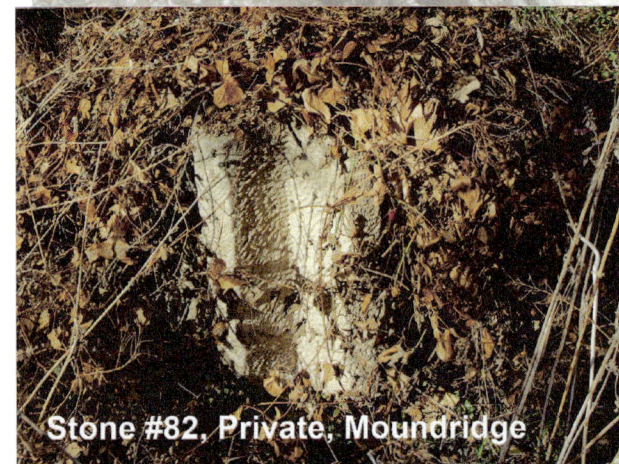
Stone #82, Private, Moundridge

These are the other three stones at the same residence as stones #75 and #76. Nothing is known of their provenance.

Threshing Stones #83 and #84

I have contacted many Mennonite associated museums across North America, including the Mennonite Heritage Village in Steinbach, Manitoba. I sent and email to the Museum Curator, Roland Sawatzky.

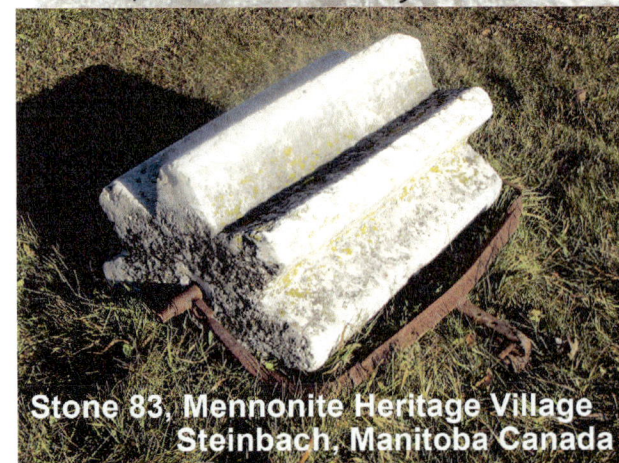
Stone 83, Mennonite Heritage Village Steinbach, Manitoba Canada

He responded saying that they actually had two stones in their museum collection, but he would have to find out more about their provenance and get back with me. After a while, I received two photographs of the stones. They were actually made in that area for demonstration purposes, but were never used for actual threshing on any farm. He can tell that they were made of local "Tyndal" limestone due to the unique gray blotches, typical of the local quarried stones.

Stone 84, Mennonite Heritage Village Steinbach, Manitoba Canada

These are both longer than most stones at 34", but both are 24" in diameter and have 7 ridges. Both have steel towing frames.

a trip to Europe, particularly South Russia. After his return, he moved to Newton in 1886 and established the Newton Milling and Elevator Co. The mill started with a daily capacity of 10 barrels and grew to a capacity of 1,700 barrels, with the flour being sold throughout the United States and Europe.[26]

Hard winter wheat or Turkey Red wheat, had been sown by the Mennonites immediately after their arrival in Kansas. It was Warkentin, however, who experimented with the various varieties of wheat in cooperation with Mark A. Carleton, the representative of the U.S. Department of Agriculture mentioned above.[27]

Initially the Turkey wheat met with slow growth and acceptance. The first settlers brought small amounts, ranging from a few pounds to as much as a bushel or two per family. A bushel was required to sow one acre. With these amounts, only small plots could be seeded, and no doubt most of each crop was needed for food and feed, rather than for seed increase alone.[28]

In the mid 1880s, Warkentin sought to correct this situation by importing 10,000 bushels of seed from Russia.[29]

However the flour industry was not quick to change, so the hard winter

Bernhard Warkentin. MLA

wheat met with slow acceptance. Millers were not well-equipped so they discounted the price. This opposition had some effect on the popularity of Turkey Red wheat. When housewives, used to the flour from soft spring wheat, tried to make bread from hard wheat flour they had even greater difficulty. Consequently, most millers rejected this Turkey wheat as unfit for milling purposes.[30]

However, there were at least two prominent reasons for the millers willing to grind Turkey wheat; it could be bought cheaper and chemical analysis indicated that it would make a flour of high gluten content. The latter quality caused a great demand for Kansas flour for export.[31]

The ability of the mills to process the wheat was only one step in the entire process of marketing the wheat. The wheat had to be transported to the East.

Halstead Mill - 1881. Kansas State Historical Society

This was the brilliance of the AT&SF in bringing the Mennonites to Kansas in the first place. Now they had a product to ship by rail back east, relieving the empty eastward runs with carloads of Kansas flour.

Farming on the Central Plains was a formidable task that had driven many farmers to desperation. The Mennonites were much better at it than most other American settlers. As is usually the habit of immigrants, they deliberately picked out the prairies of Kansas and Nebraska as having land and climate similar to that which they left behind in their homeland. They had experience in farming on prairie land and had seed that was well-adapted to the harsh Great Plains environment. It was the Mennonites who got the gigantic wheat empire of the Great Plains underway.[32]

The previously mentioned radio drama concluded, "Where Kansas had not produced as much as 5 million bushels of wheat in any one year prior to 1874, after the Mennonites arrived with their hard winter wheat,

Kansas soon became the leading wheat producing state in America...the Kansas wheat is high in quality and commands a premium on the markets of the world. Many eastern bakers now use no flour except that milled from Kansas hard winter wheat."[33]

As a prominent European flour merchant said a few years ago, "American wheat stands high in the estimation of the people, but Kansas leads the world. There's nothing like Kansas wheat; no other with so much gluten, so much muscle and nerve-building material in it. It has become so popular on the Continent as well as in Great Britain that our dealer has to carry it, regardless of price or profit."[34]

In 1974, 100 years after the Mennonites arrived in Kansas, the State of Kansas created a Wheat Centennial Board to develop ways to commemorate the arrival of Turkey Red wheat in Kansas. The committee co-chairs were Harley J. Stucky and Olga Krehbiel Stucky. Celebrations were enjoyed all over the state.

Kansas Highway 50 was declared the "Turkey Red Highway," commemorated with road signs. The big event of the year was the Ceremony of Kansas Hard Winter Wheat Stamp program at Tabor College. The Postmaster General of the United States Postal Service, E. T. Klassen, gave the address when the "Rural America - Kansas Hard Winter Wheat" stamp was released.

Goessel and other towns also had centennial celebrations. Goessel contin-

Grain Sack - 1865 - From the village Leske in West Prussia. KM

Stone 85, State Historical Society of North Dakota

Threshing Stone #85:

I found this stone via contact with the organization Germans from Russia Heritage Society in Bismark, North Dakota. I contacted Michael Miller who forwarded my request to others, of which I was contacted by Mark J. Halvorson, Curator of Collections Research at the State Historical Society of North Dakota.

He sent me the following information on a threshing stone in their collection, along with images. The stone is roughly 4' long and 18" in diameter, made of granite. It is associated with a Russian family from near

Benedict, North Dakota. Leonty Teachmond (1856-1925) emigrated from the area of Odessa to the United States in about 1899. The stone was acquired by The State Historical Society of North Dakota in 1953.

Threshing Stone #86:

I had been tracking this stone for well over a year before I finally found it. I was first told by Brian Stucky that there once was a threshing stone on the grounds of the Kaw Mission in Council Grove but it was not there the last time he had visited. He remembered it setting on a concrete pad and that the stone was slightly smaller than what we are used to seeing.

Ray Wiebe also recalled seeing a threshing stone there with a plaque that said "Mennonite Stone Thresher." I was also told by Harry Kasitz, who used to live near Council Grove, that one used to be at the mission but it was no longer there.

I sent two emails to the museum but never received a response.

While on the MCC Sale Motorcycle Ride, we visited the McPherson Mennonite Church

ues to celebrate with the Threshing Days every year. By 1974, Goessel had built the Heritage Museum to preserve artifacts from the early Mennonite families who settled in the Goessel area. The museum complex now includes the Turkey Red Wheat Palace.

There were even plans for a Turkey Red Wheat Museum to be located in North Newton, to showcase the legacy of wheat in Kansas. The building was to be designed in the shape of a threshing stone, but this vision was never realized. (See architectural drawings in the Symbolism Article.)

Many articles, books and pamphlets were created to document this centennial. *The Wichita Eagle* published an article by Joanna K. Wiebe titled, *"Turkey Red Wheat Centennial Plans Ripening for 1974,"* telling about the history of Turkey Red wheat in Kansas to a whole new audience that knew little of this story.

By now, newer varieties have replaced the old Turkey Red Wheat. But in 1974, some farmers planted small fields of Turkey Red wheat for centennial celebrations. I painted a road sign that was placed in my father's wheat field saying "This field is Turkey Red wheat com-

Kansas Wheat Centennial 1974 - State of Kansas named Highway 50 "Turkey Red Highway." Florence, Kansas.

memorating 100 years in Kansas."

No doubt, Kansas has played a significant part in the world's food supply. Many can be proud of the legacy that ultimately made Kansas the Breadbasket of the World.

Currently the Kansas Department of Agriculture says, "On average, Kansas produces more wheat than any other state. Nearly one-fifth of all wheat grown in the United States is grown in Kansas. Kansas ranks first in the nation in flour milling, wheat gluten production and wheat stored. Roughly one-third of Kansas' 63,000 farmers grow wheat. Normally, Kansas farmers produce about 400 million bushels of wheat a year, with a production value that hovers around $1 billion."[35]

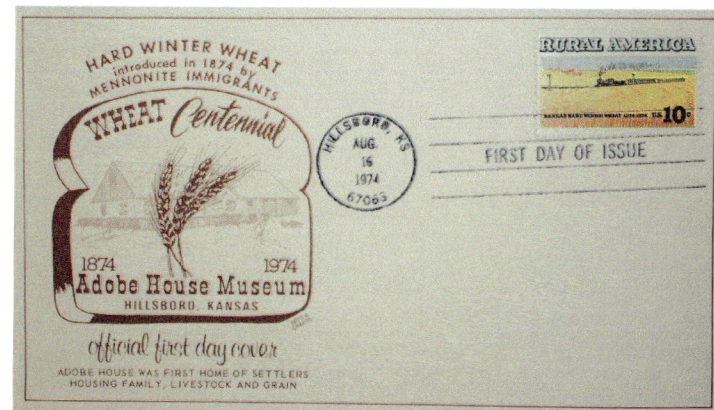

Kansas Wheat Centennial First Day Issue Stamp "Rural America, Kansas Hard Winter Wheat 1874-1974." Author's collection, gift from Thelma Bartel

05.20.2012 16:04

Symbolism & Bethel College

Humans are unique in both their intellectual capacity and their desire to use symbolism in objects, actions, or ideas. Symbols have been used throughout history to represent the meaning or memory of something other than itself. It shows up in all world religions, governments, organizations, families and even as personal ritual. These symbols assist us in heightening experiences by making an abstract concept into a more tangible experience.

Rocks and stones alone have been used to symbolize strength and tradition. As early as 500 BCE, large stones were erected across prehistoric Europe to stand in lines or in circles, such as Stonehenge in England. However, their absolute meaning is unknown. In Buddhist symbolism, water and stone are the Yin-Yang, symbolizing that opposing forces are interconnected. The Bible is full of stories about stones: Jacob used a stone as a pillow, Joshua set up memorial stones, David chose five smooth stones to fight Goliath, Elijah took twelve stones to build an altar, and Living Stone is a metaphor and a paradox suggesting strength and life simultaneously.

Three world religions are all focused upon a particular rock in their scriptures. The "Rock," now referred to as the "Dome of the Rock," is used to symbolize a variety of things. It is reputed to be Mount Zion, the location of a Jebusite fortress.

Prior to King David it was a community threshing floor on which either grain was threshed or a rock upon which the grain was beaten to thresh out the seeds. According to Jewish Law, Mosaic sacrifices can only take place on that one rock.

In the Bible, II Kings and I Chronicles tell the story of how King David purchased the threshing floor and built an altar there.

That same rock is considered by many Jews and Christians to be Mount Moriah, the site where Abraham almost sacrificed his son, Isaac. This rock was in the Temple Mount and part of the "Holy of Holies," the inner sanctuary of the Temple in Jerusalem, the most sacred site in traditional Judaism, where the Ark of the Covenant was kept during the First Temple.

The "Dome of the Rock" is now a Muslim temple built over that same "threshing floor." For Muslims, the rock was sanctified by the story of the Prophet Mohammed's Mi'-raj, or Night Journey, to Jerusalem and back to Mecca (Qur'an 17:1). From the top of the rock, Mohammed began his ascent to heaven. Today, the Mosque of Al Aksa and the Dome of

Opposite - "Touching the Stone" at Bethel College graduation ceremonies - Seniors Alyssa Schrag, Carrie Schulz, Nate Snyder, Megan Schrock, Allison Schrag - 2012.

Dome of the Rock. Library of Congress

the Rock sit on the Temple Mount, marking the third holiest site in Islam. This former "threshing stone" has become a true symbol for holy worship

In Biblical writings, threshing and the threshing floor are used as a metaphor for judgment, speaking of the process of separating the good from the bad, especially in the prophets (Dan 2:35, Isa 21:10, Jer 15:7, 51:2, Hos 13:3). Then, in the New Testament, John the Baptist used this metaphor as a warning of the activity of the Coming One (Matt 3:12, Luke 3:17). There are actually 36 references in the Bible to the threshing floor.

Jesus is the solid rock upon which to build our lives (Matt 7:21-27), a theme in New Testament writings. Jesus also used rock metaphors in his parables, such as a wise man building his house upon the rock (Matt 7:24).

In addition to the symbolism of stones and rocks in religious literature, there are many references to the tools and practices of ancient farming. Frequently analogies are drawn between these farming images and spiritual/religious truths. References refer to flails or sticks which were used for threshing (Ruth 2:17; Isa 28:27) and there was also a threshing instrument which was drawn over the grain, which the Hebrews call the *moreg*, a threshing roller or sledge, which was something like the Roman *tribulum* (Ruth 2:17; Isa 28:27; Isa 41:15, 3:15; Amos 1:3, 2 Sam 24:22, 1 Chr 21:23).

In the Bible, the soil was prepared, the seed was sown, broadcast over the fields, and a harrow was mentioned (Job 39:10). The reaping of the grain was cut and put in sheaves (Gen 37:7; Lev 23:10; Ruth 2:7; 15; Job 24:10; Jer 9:22 and Micah 4:12).

Jesus also used analogies of farming in many of his parables, including, one where the kingdom of heaven is like a man who sowed good seed in his field which landed on different types of soils (Matt 13:24-30).

Needless to say, after all those references from various religions, the processes of farming and threshing provide a wealth of symbolism for us to use.

and enjoyed a potluck lunch in the basement. My brother LaVon who attends that church had been asking around if anyone there knew of any threshing stones.

While getting food I met Stan Miller who asked me if I knew that there was one on display at the Kansas History Museum in Topeka in an exhibit called "150 Things I Like About Kansas." I said I was not but I would check it out.

I sent an email to Kansas State Historical Society, but got no response. I asked John Thiesen who I could contact at KSHS and he gave me the name Blair Tarr - Museum Curator. I sent him an email about my quest.

He responded with detailed information on the stone. It was originally used in the Piazzek Hill in Grasshopper Falls on the Delaware River, and it was indeed in the museum gallery. He continued on telling of another stone at the Kaw Mission Site Historic Site at Council Grove. The provenance for it is as follows: "Mill stone used in Saunders Flour Mill which began operation in Council Grove in 1874."

Had I also found the Kaw Mission stone, or was he confusing a mill stone with a threshing stone? I was discouraged.

I sent him an email back questioning his comments and also attached a picture of a typical "Threshing" stone for his reference. He responded with apologies for the mix-up. However he continued with, "We do have a threshing stone - 1980.96.544 - and it has been on exhibit since January.

I went to the museum and got a photograph of the stone on exhibit and spent time in their archives doing stone research. I met with Mr. Tarr and he gave me the KSHS Object Worksheet for Accession # 1980.96.544: "Date Range 1860-1880 - Description: Stone wheat thresher roller with seven rounded ridges extending across width. Iron rod extends through center - Provenance: According to label, "This stone thresher was made and used in Morris County in the 1860s by German-Russian immigrants. Horses pulled the stone over the

GRAY MAROONS.

In all parts of the world where threshing was done, the threshing floor was a community gathering space, not only devoted to threshing but a center where important community events took place and rituals were celebrated.

Symbolism has its place in our time too. Symbolism is not lost at Bethel College. As discussed in the "Finding the Stones" dialogue (through the center of this book), the threshing stone has had a long-standing relationship with the college. The first threshing stone was brought to the campus in 1903 by the first president, C. H. Wedel, to stand in front of his home. The *Bethel College Monthly* of June 1903 noted, "Professor C. H. Wedel recently acquired an interesting relic in the shape of an old threshing stone, such as were formerly in general use in Russia . . . this valuable relic of former agricultural methods now occupies a prominent place in Professor Wedel's yard, but there is little danger that anyone will carry it away."[1]

Eventually a second stone came into the possession of Wedel; then, later these two stones found their way to the front of the Science Hall.

On November 16, 1934, the threshing stone was adopted as the official symbol of Bethel College.[2] According to John Thiesen's research, presented at a faculty lecture in 2007 titled "*What's a Thresher*", "this was sort of a mysterious decision, like many others. It's not mentioned in faculty or board minutes; there were no news releases. Presumably it was the decision of Edmund G. Kaufman.

The "Collegian" reported on a brief ceremony, where two threshing stones, one of them had been in C. H. Wedel's yard, were placed in front of the old Science Hall, on either side of the main door. Apparently this started a new tradition for the college, an annual freshman initiation ceremony at the threshing stones. The very limited reporting about the new symbol made explicit reference to

the "pioneering spirit" and all those clichés.[3]

The 1936 yearbook, the *Graymaroon*, showed images of the ceremony and contained several graphics inspired by the threshing stone. Stickers were printed with the new symbol and given out to students.

Text in the *Graymaroon*, says: "This threshing-stone, hewn out of solid rock, has been chosen as the emblem of Bethel College. It symbolizes certain fundamental characteristics of the school: The pioneering spirit, simplicity of life, faith and stability, solidity of character.

"A year ago almost three hundred students shivered around the Science Hall steps to adopt this homely implement as the emblem of Bethel College. In spite of the chilly November air, three hundred hearts warmed to Mose Stucky's words as the student body dedicated the emblem.

BETHEL COLLEGE GOLDEN ANNIVERSARY

FAITH — HOPE

SEEK YE FIRST

THE KINGDOM OF GOD

LOVE

NEWTON October 12 1938 KANSAS

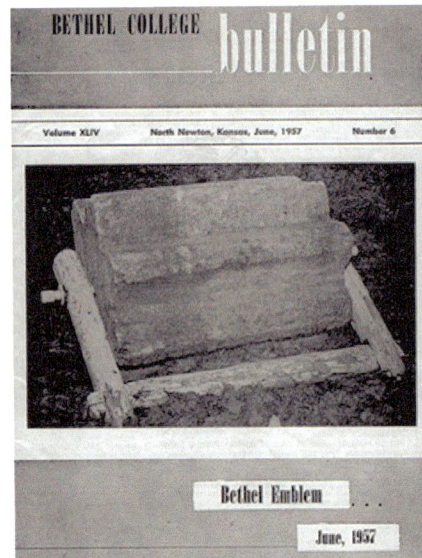

"Bethel College Bulletin" - 1957. MLA

Half a century ago one gray stone building represented the hopes for a Mennonite education. The men who built this school were primarily farmers. To thresh their wheat, they designed a threshing stone, the origin of our school emblem. That rugged, sturdy stone typifies the simple, hardy spirit of the pioneers who used it."[4]

In 1938, the image of a threshing stone in use in front of the Administration Building was incorporated into the official Bethel College Golden Anniversary seal.

In the following years, the threshing stone showed up in letterheads and evolved in various uses throughout the college's printed materials. The threshing stone was not always depicted graphically correctly, sometimes appearing with as many as twelve ribs and as few as six teeth, with varied depths and tooth designs.

Stone 86, Kansas Museum of History
Topeka, Kansas

wheat and bumped the grain much as if it had been flailed. Stone thresher, introduced into Kansas along with the Turkey Red Wheat by the Mennonites (Russian/German settlers) who settled in McPherson and Harvey Counties in the 1870's. Found at Kaw Mission site with no accession number, February 1980."

When I read this I was excited to have finally found the lost Kaw Mission stone. In further talking to Mr. Tarr, I asked him some probing questions, and he volunteered to do some further digging.

After about an hour he found the accession paper from the Kaw Mission Museum, which contained some new information. Name of Donor: Herold Anker - Address: Wilsey, Kansas.

This stone is smaller than most stones, (28" long and about 20" in diameter). It is more crudely cut and not all the ribs are evenly spaced. There are no tell-tale chisel marks and the stone has a smooth finish, with numerous chunks knocked off.

The iron axle is only partly into the end of the stone, wedged in a square hole about 3" deep with steel shims.

Threshing Stone #87

This stone was brought to my attention by John Thiesen. As stated earlier, this is one of the other stones from the Thiesen family and it is still located on the original farmstead. I was able to contact the owner's

Stone #87, Private, Goessel

Over the next twenty years, the imagery showed up on many things from the back of letter jackets to homecoming floats.

The June 1957 issue of the *Bethel College Bulletin* featured the threshing stone prominently: "This is a Threshing Stone . . . to Bethel College this Threshing Stone Emblem symbolizes:
1. The Pioneering Spirit
2. The Simple Life
3. Faith
4. Stability of Character
5. Hard Work
6. Co-operation
7. Brotherhood."[5]

In the 1960s Robert Regier had an opportunity to design a new Bethel College logo. He used the "BC" graphic to represent stacked books as well as an abstract view of the ribs of a threshing stone from straight on.

Annual Report of the President

Robert Regier Logo - circa 1966. MLA

The redesign of the Bethel College logo led designer Ken Hiebert to reflect, "One of the first questions to come up was the fate of the threshing stone as symbol. Younger generations have little if any connection to the archaic tool; as literal image, it did give athletics an image that had to do with threshing in the sense of winning. But the historical meaning, which coupled threshing and wheat with the pioneering spirit of the founders, was a story that was no longer self-evident.

"I raised the question about the continuing validity of the threshing stone as a college symbol with John Janzen '61, with whom I had worked on the Kauffman Museum logo. From an anthropologist's viewpoint, he extolled the universal qualities of both threshing and the fundamental material of stone as belonging to nearly every early culture. He thought it was too valuable to discard.

"My next attempts at visualization were then to find if there was a way to use the threshing stone as a base, but giving it a form coming from its symbolic meaning of discernment, the idea contained in the separating of wheat from chaff. For me this seemed to be the ongoing validity of the symbol, relevant to every departmental pursuit of the college. I wanted to somehow give it a resonance and energy that could be sensed without knowing the background, but which would have an additional layer of meaning when this was known.

"Out of these studies came the 'legacy symbol,' as we call it. The diversity of the angular forms is brought into harmony by the 7-pointed, star-like figure of the stone in profile. The centrifugal force of the curved lines gives it motion. The wheat curve culminates in a kernel, a seed which recapitulates the home form of the threshing stone, ready to be reborn."[6]

The Bethel College web-site continues to quote Ken where he so eloquently expresses the underlying meanings further, "As the stone is usually presented it has a rather forlorn look, unmoving, its yoke lying disconnected from the source of its motion. Its essential meaning, however, is dynamic—historically as a symbol of the agricultural transformation of the plains—but more transcendently or timelessly, as a metaphor for the separation of the grain (valuable inner) from the hull (outer). As such it also has to do with the spirit of discernment. In this sense, the threshing stone is a humble reminder and symbol of the sorting process that is education, here given a connection to the whole teaching and learning enterprise, and more relevant than ever in a world of information overload. Threshers are thus not just playing volleyball or running the relay; they are scientists, musicians, poets, nurses, social workers—in short, persons of any discipline—each bringing a powerful value system to bear on their professional development. . . . The threshing stone honors our legacy and reminds us of the key qualities that shape leadership — moral, spiritual and professional."[7]

The threshing stone is hopefully the symbol of Bethel College forever, even as it is used and abused in many ways, or continues to get ridiculed. It even shows up on a former blog called "Rocket Queen," discussing strange college mascots. The "Threshers" made the "Weirdest Mascot" list.

Yet this icon endures. A revival of sorts hit campus in about 2010, where many cool designs and applications of the symbol are applied to everything. This updating of the imagery has seemed to be a hit with students and alumni, with T-shirts saying, "I'm proud to be a piece of old farm equipment" and "Rolling Stones."

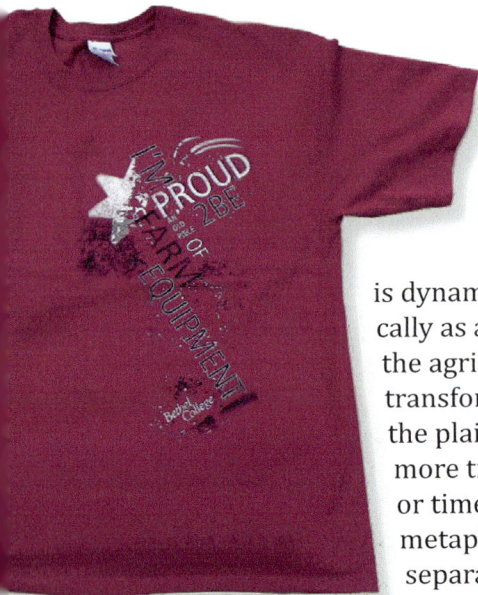

Bethel College shirt: "I'm Proud to be a Piece of Old Farm Equipment." Author's collection

son who gave me permission to photograph it. This stone was displayed in the yard on its side.

Threshing Stone #88:

I had known about this stone for quite some time, but it got lost in my notes. I had heard about it from Johnny Schroeder at the Goessel Threshing Days. I called the owner Eugene Franzen, and he confirmed that he had a stone and I could come photograph it.

Stone #88, Private, Goessel

I knocked on the door and startled the dog inside the house and soon Eugene came to the door. He said, "Ya the stone is out back next to the threshing machine," and that I could go photograph it, but he and the dog would stay on the porch, and to be careful crossing the electric fence into the goat pasture.

Making my way back through the pasture I found a threshing machine, but no stone. I kept looking in larger and larger circles around countless old pieces of abandoned farm equipment till I spotted another threshing machine under the hedge. Next to it, on its side with a long axle through it. It was a nice threshing stone.

I got several photographs and went back to the house. I had a nice long talk with Eugene. He had been friends with my in-laws, and Karen's dad's home place was just down the road a ways. He said this stone was not originally from this farm but he and his brother had bought it at a farm sale near Buhler about 50 years earlier. Eugene died just several weeks after our meeting.

Threshing Stone #89

This stone is a famous stone that virtually no one remembered, but fortunately it finally found me. Earlier in my project I had stopped to talk with Marci Penner, the found-

"A New Tradition at Bethel"

First-year freshmen at Bethel College continue a tradition begun in fall 2006. "Touching the Stone" has become a rite of passage for freshmen students as they come to Bethel and attend their first convocation. This tradition has also been added to the commencement ceremony. As the bell is rung and the graduating class enters Thresher Stadium, two lines of students pass by the old threshing stone, "Touching the Stone" as they end their time at Bethel.

The symbolisms are almost endless when talking of threshing, rocks, wheat, chaff, and Bethel. The symbolism is strong, steeped in world history, Mennonite history, and Bethel history...the threshing stone is still a great symbol.

The threshing stone in front of the Bethel College Administration Building - 2012.

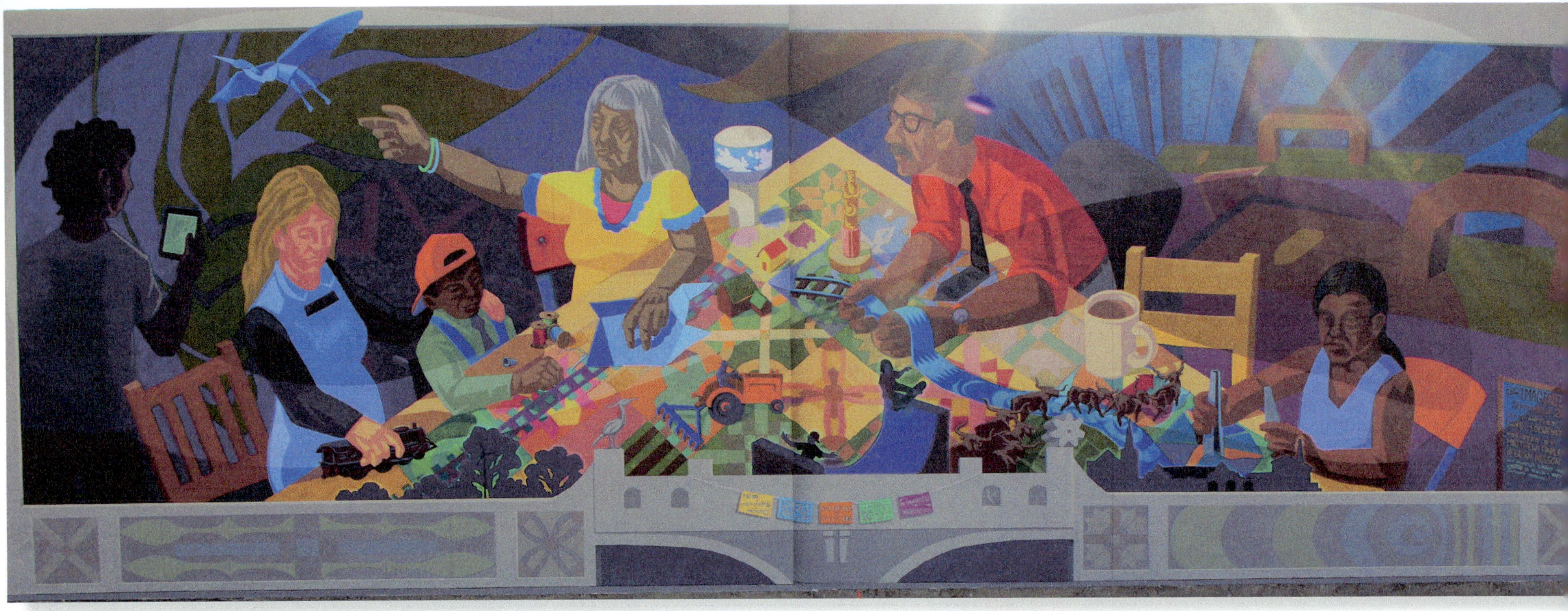

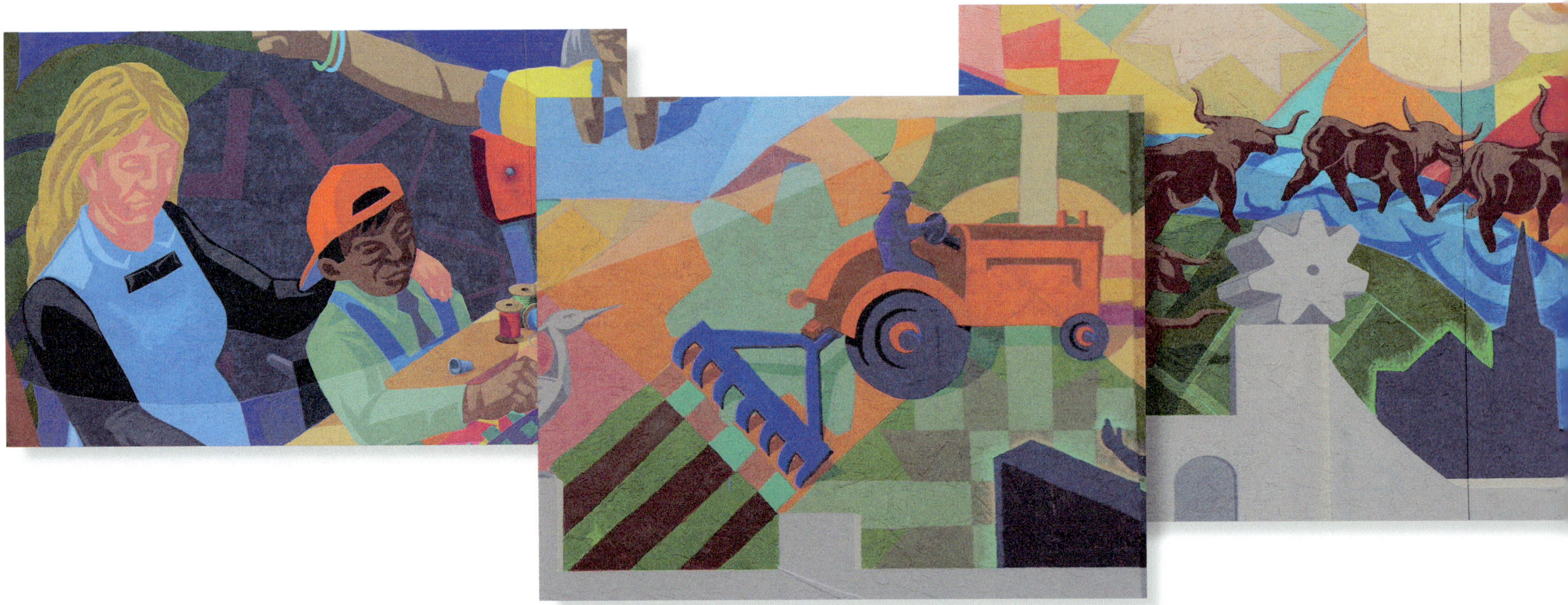

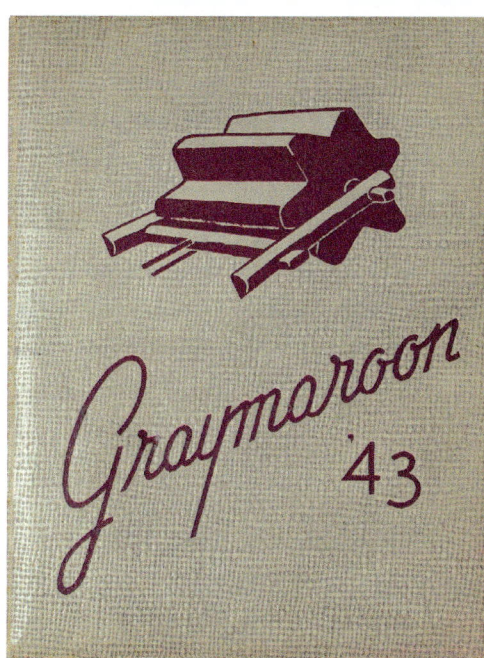

Bethel "Graymaroon" 1943. Author's collection

Wheat Centennial plate.
Author's collection

Opposite - "The Imagineers." Downtown Newton, Kansas mural depicting Newton characteristics by David Loewenstein - Image includes 3 threshing stones - 1. Bethel College logo, 2. Behind the tractor, 3. On a pedestal. Courtesy David Loewenstein

Art and Objects

The threshing stone has been inspiration for many artists and craftsmen, showing up in art and objects that bear the unique angular shape of the stone. As art, it can be seen in paintings and sculptures. In graphics and objects, it has been reproduced in many forms, many of which are associated with Bethel College, where this ancient farming artifact is used as the mascot of the college.

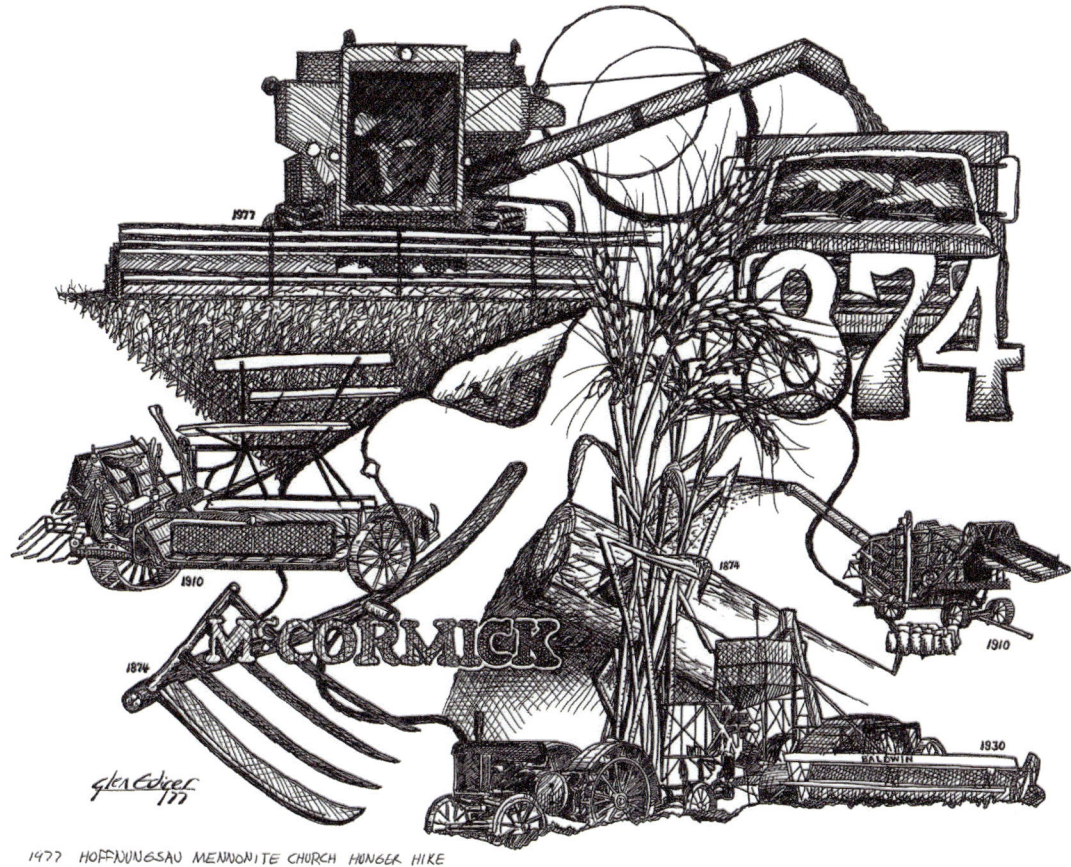

Drawing done in 1977 by Glen Ediger for the Hoffnungsau Centennial Celebration, depicting wheat farming history.
Note the threshing stone in the middle; the round tooth stone was drawn from the threshing stone in front of the Bethel College Administration Building - Original series was 100 lithographs.

Bethel College mug.
Alumni Office

Flying Disc.
BC Alumni
Office

Painting "Progress of Agriculture" by P. H.
Friesen - 1974 - One of a series of large
paintings. MH&AM, Goessel

Steel line art - Wheat Palace. MH&AM, Goessel

Desk set given to Harley Stucky in 1974 in recognition for his co-chair position for the Kansas Wheat
Centennial. Author's collection from Frank Stucky

ing Director of the Kansas Sampler Foundation and author of
numerous books about Kansas.
I have known her since we were
young; we grew up in the same
church. I wanted to talk to her
and see if she knew where any
threshing stones might be. She
has traveled to every one of the
627 cities in Kansas and virtually every square mile in between
and probably knows the state
as well as anyone ever has. I did
not want to miss access to any
of her knowledge or contacts.
She wrote about my project and
the story was noticed by one of
her readers.

I was contacted by Ann
Birney, a well known historical re-enactor in the state of
Kansas with information about a
new stone find. "My mom called
after reading of your quest. She
reminded me that my aunt has
my grandfather's stone...because
it is 'published' you will want to
know about it. It can be seen
on p. 111 of Bliss Isely and W.
M. Richards' The Story of Kansas, Topeka: Kansas State Print-
er, 1958 (the pink book that
my generation used in grade
school).

"Marvin Richards was my grandfather, then the superintendent of Emporia schools, and passionate about history. Unfortunately, the origins of the stone are not known to my mother or her sister, in whose backyard the stone now sits. Aunt Millie has a photograph of her with the stone in front of Emporia Junior High when she was, she believes, age three, putting it at about 1952. She says that someone gave it to Daddy, and, because, as Mother says, Daddy was instructional, the stone went to the school where it could be discussed. When Grandfather retired the stone went to their home where I remember it in his garden with a bird bath. Long after Grandpa died the stone went with Millie and Ron Felkner to their various homes in Kansas City. So there you have it—a Kansas threshing stone much moved!"

I had previously seen the picture in the Isely/Richards Kansas history books at several libraries and even bought a used copy for my own collection, but I never expected that I would actually find the real stone someday.

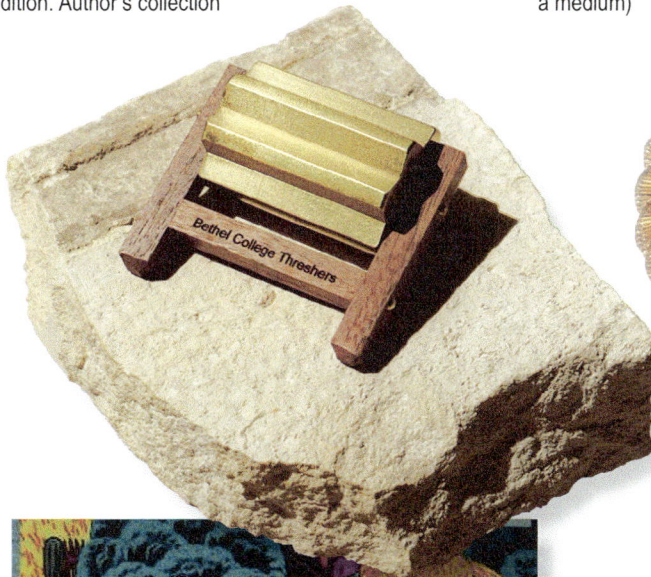

Bethel College 125 Anniversary Collectible, engraved brass and wood model on stone base from the first step of the Bethel College Administration Building - 75/125 limited edition. Author's collection

Wheat Weaving by Adelia (Jantz) Stucky. circa 1976 (No threshing stone, but a beautiful use of wheat as a medium)

Chinese painting - Xian Threshing. Allibaba.com

Carving in wooden chair by Chet Cale. Courtesy Chet Cale

The meeting day evolved into a virtual family reunion with Ann, Joyce, Ann's mother Betty Dunbaupt, and her aunt Millie, along with several other family members. It was quite an occasion, with lots of excited discussion. (I think every family loves to be the center of attention.)

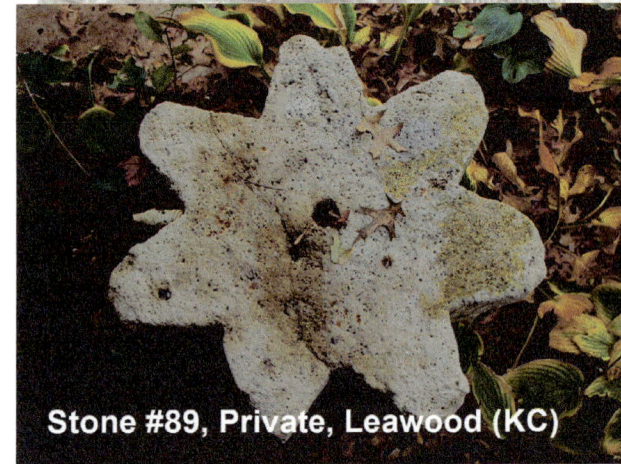

Marie and Martha Voth wheat marquetry - 2012.

Buhler, Kansas city sign repainted in 2012 with threshing stone.

Stone #89, Private, Leawood (KC)

Bethel College Homecoming float - 1963. MLA

"Mennonite Settler" by Max Nixon - 1942 - Commemorating the introduction of Turkey Red hard wheat by Mennonites in 1874, helping make Kansas the Breadbasket of the World. (Possibly the base was inspired by the threshing stone)

We finally made it to the back yard to shoot some pictures, and I recreated Millie standing next to the stone from the same angle, lining up some of the distinguishing features of the stone, with her hand resting on the stone. One of the jokes of the day was "How much the stone had shrunk over the years."

Millie Richards standing by the stone in 1952.

FOUR CENTURIES IN KANSAS

STONE THRESHING MACHINE

The stone was notched by the Mennonite farmer of Harvey County who owned it. It was drawn back and forth across the wheat straw to beat the grain from the heads. The farmer who made it later gave it to the schools of Emporia and it stands on the junior high school grounds.

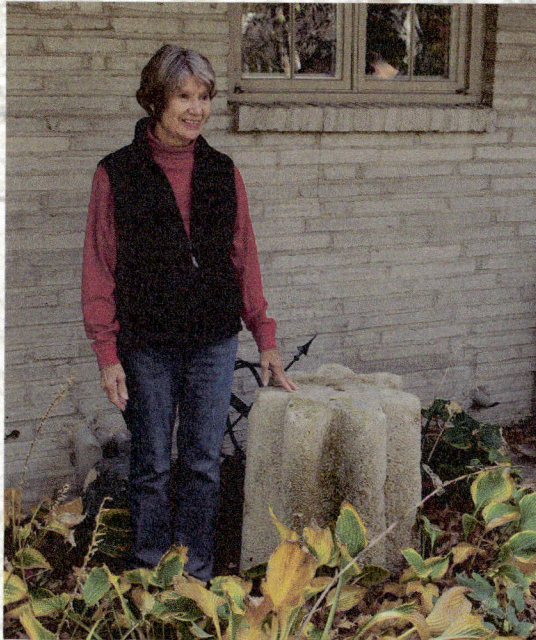

"Four Centuries in Kansas" -"The stone was notched by the Mennonite farmer of Harvey County who owned it. It was drawn back and forth across the wheat straw to beat the grain from the heads, The farmer who made it later gave it to the schools of Emporia and it stands on the junior high school grounds."

Millie Richards standing by the same stone 2011.

Logo of MH&AM, Goessel.

Memorial In Neukirch, Ukraine. The school director, his assistant, and the history teacher have developed a memorial for victims in seven neighboring villages that includes 8 threshing stones. Courtesy Ben Stobbe

Threshing stone reproduction - by Galen Froese on South Main Street, Buhler, Kansas.

Monument in memory of the Mennonite settlers who founded the fifty-seven Molotschna Mennonite villages in the years after 1804 -by Paul Epp - 2004 - "Mennonites in Ukraine were identified by their agriculture...what better symbol of Mennonite agriculture than a threshing stone."
Courtesy Nathan Dick

Mennonite Centennial Memorial, 1974, Peabody Kansas - by Solomon Loewen. Includes an actual threshing stone cut into 4 pieces.

Wheat Heritage Engine and Threshing Co. Logo, Goessel. Author's collection

Bethel College Letterhead detail - 1941. MLA

Print by Robert Regier - 1966. MLA

Wooden Recipe Holder - 2010. Author's collection, gift from Arlin Buller.

Goessel Mennonite Centennial Plate - 1974. Author's collection

Mennonite Threshing Stone. Many of these cog-shaped stones were hewed from solid limestone by the Mennonites in central Kansas in 1874. Wheat was placed upon the ground and the threshing stone in a horizontal position was pulled over the wheat to separate the grain from the straw.

"The Story of Kansas" 1953

Photo by Herbert Schmidt.

Stone thresher used by the Mennonites to thresh the first Turkey Red wheat in Kansas. The star shaped picture is the end view of the stone. Other picture is front view. The stone was rolled over the wheat and in that way beat out the grain.

"Early Days in Kansas" 1967

Art made with wheat grain - by Marie and Martha Voth - 1974. MH&AM, Goessel

Bethel College License Plate. BC Alumni Office

However, we still have a mystery! The stone in the 1967 edition of the Kansas history book is a square toothed stone not the round toothed Richards stone, and the photo credit was from Herbert Schmidt. So where is this stone? It is very possible that it is stone #34.

Threshing Stone #90

This stone came to my attention when I had a table display set up at the Inman Santa Fe Days. I was told that a man west of town might have a threshing stone on his farm, but they said he already had a farm sale and it was most likely sold. So that afternoon I headed out to his place. I pulled up to the farm and saw that it was pretty well all cleaned up and saw little hope that he would

Stone 90, Private, Windom, Kansas

Liberty Bell made of woven wheat in 1976 at the request of the Smithsonian Institution. MH&AM, Goessel - Photo courtesy Dan Bergen

T-Box Marker for Bethel College Golf Tournaments - Stone, Arlin Buller - Yoke, Rod Schmidt - Laser inscription, Al Penner. Courtesy Arlin Buller

Bethel College Hoodie. Author's collection

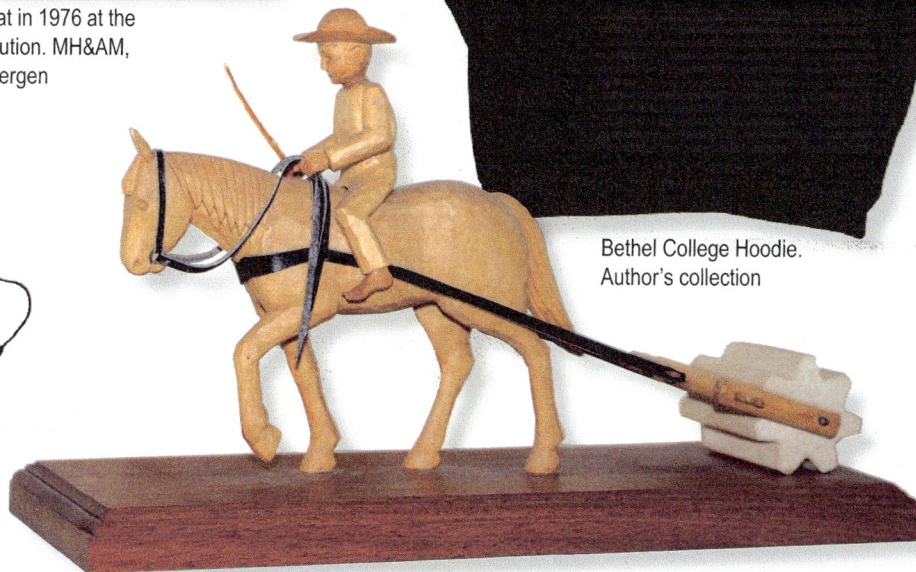
Stoney Power - by Jess Rempel - 2006. MLA

Model carved by Richard Schmidt - One of 3 known. One was exhibited at the Smithsonian, Division of Work and Industry in 1975 (object 76A09). MH&AM, Goessel

Concrete reproduction threshing stone by Rod Funk. He made about 20, they are actual size.
Judy Friesen

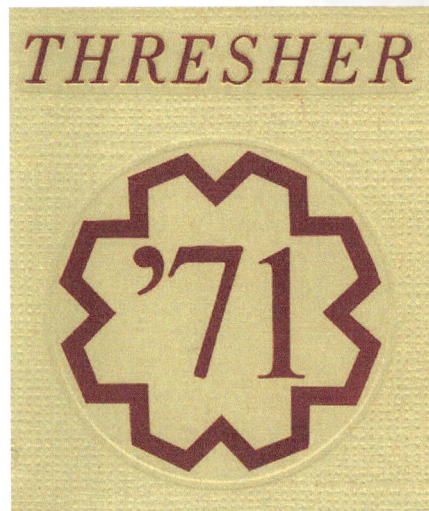
Threshing carvings. by Jacob J. Janzen - circa 1970 MH&AM, Goessel

THRESHER '71

"Thresher" Cover - 1971. MLA

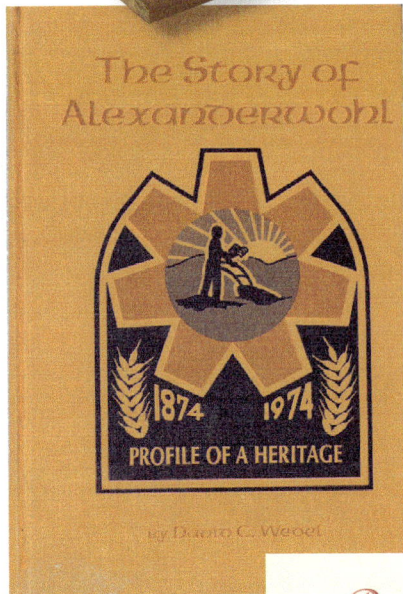
The Story of Alexanderwohl

1874 1974

PROFILE OF A HERITAGE

by David C. Wedel

"The Story of Alexanderwohl"- Book cover design by Mennonite Press - 1974. Author's collection

BETHEL COLLEGE THRESHERS

Emblem by Robert Regier - 1966. MLA

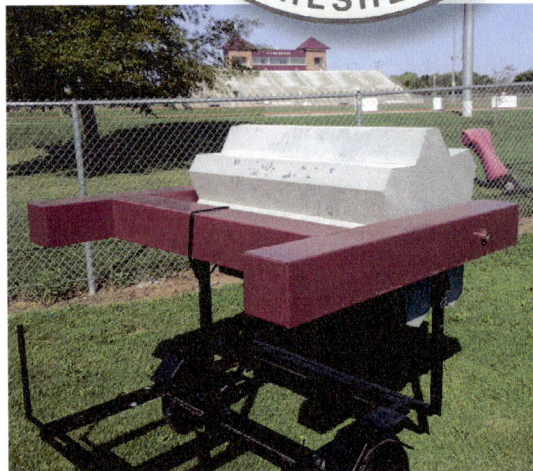
Aluminum Thresher mascot, double size, 2009. Bethel College Athletic Department

Bethel College '56

Commencement Invitation - 1956. MLA

still have it, but I drove onto the yard and there it was. It was displayed on an old piece of wood, cracked in half, with a cable tied around its center to keep it together. I knocked on the door and a middle-aged man came to the door and said he was the son of the owner.

I was invited in, but we had a challenging conversation as he at age 94 and his 97 year-old friend were very hard of hearing. Anyway, turns out that the older gentleman was the former owner of the threshing stone. At one point he said he wanted to throw the old stone in the river to get rid of it but fortunately, this farmer saved it. He moved it to his farm and it has been there ever since.

The origin of this stone has no known Mennonite legacy but has always been in the Lutheran community. The dimensions and design are identical to most all found stones but the cut marks are a bit different style.

One of his other neighbors has requested to inherit it some day.

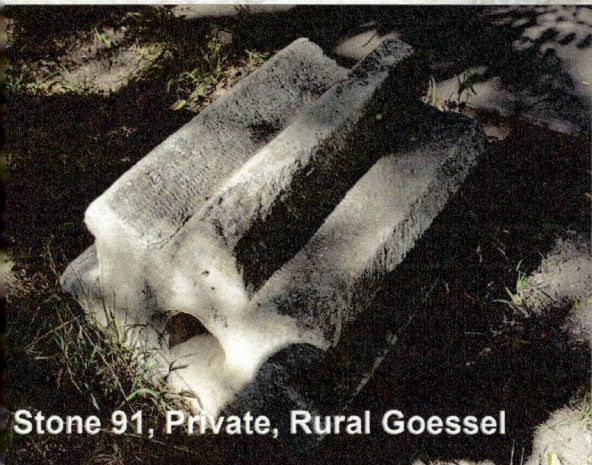

Stone 91, Private, Rural Goessel

Threshing Stones #91 and #92

This was another lead from Johnny Schroeder at the Goessel Threshing Days in 2011. I met Walden Duerksen who had two threshing stones, so I made arrangements to photograph them.

One was on the yard, near the house. He had bought

Stone 92, Private, Rural Goessel

"Graymaroon" - 1936. MLA

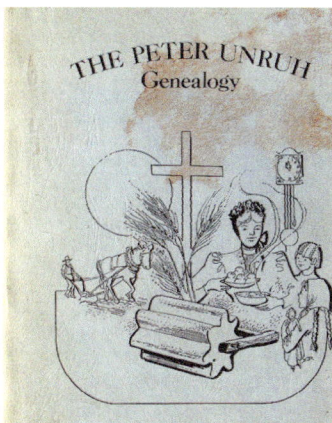

"The Peter Unruh Genealogy" - 1981 - Family Publication. Author's Collection

Stoney waving. MLA

Bethel College Women's Association replica stone about 15 inches wide. Bethel College Bookstore

Man Shall Not Live

Bethel College Brochure for the Fine Arts Center - 1963. MLA

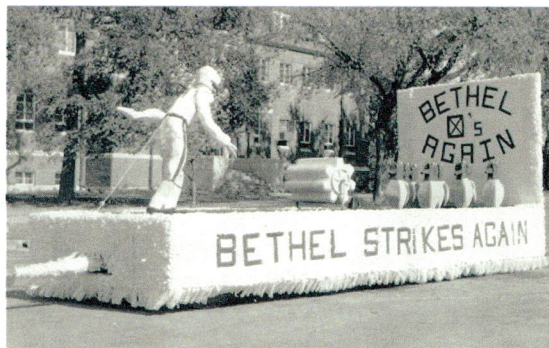

Bethel College Homecoming float - 1964. MLA

Bethel College Stress Stone, 3 inches wide. Author's Collection

Thresher Award by Virgil Penner. MLA

Stoney Graduating. MLA

Folk Festival Poster - 1960. MLA

Thresher Head foam hat worn by Bethel pep-band - 2001. MLA

Detail of Bethel College Symbol. MLA

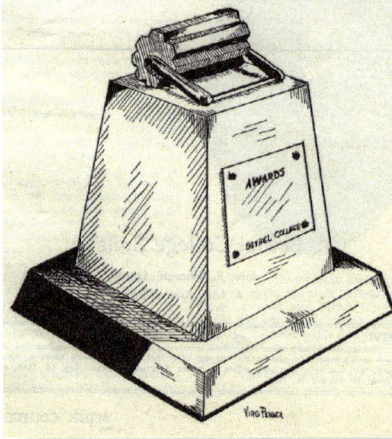

Letter Jackets - 1936. MLA

it at the A· J· Schroeder auction many years ago· It was his mother-in-law's (Schmidt) family stone· The farm was part of the Hochfeld village· The stone has a large axle hole that appears to be worn with use·

The other stone is from his great-grandpa H· S· Duerksen· It was on the Elmer Duerksen farm for many years, Elmer gave it to Walden·

Threshing Stone #93

I got a lead that Verdan Harms, of Hillsboro, Kansas, might have a threshing stone· I called him up and made arrangements to visit him· He has been involved in Hillsboro local history for many years and gave me a lot of information· He bought a new home in Hillsboro in 1963,

Stone 93, Private, Hillsboro

with the threshing stone in the back yard that was sold with the house.

This stone is now in the collection of Kelly Harms. He has it nicely displayed in his back yard. This stone has no axle hole.

Threshing Stone #94

Stone #94, Private, Newton, Kansas

This stone was another lead from Kelly Harms, actually from early on in my research, but got lost in my notes. It is originally from the Hillsboro area, but now resides on the south side of Newton, Kansas. It has rounded teeth, often seen in the Hillsboro community.

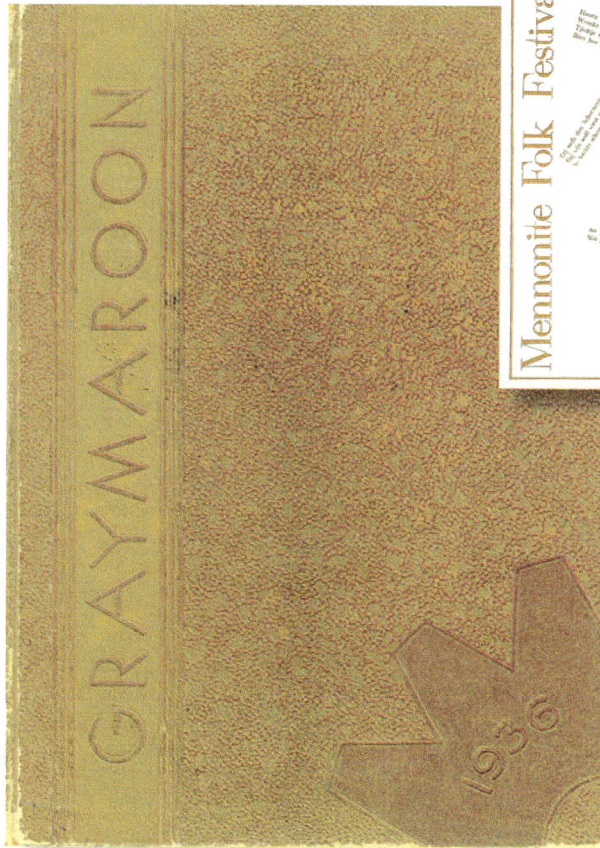

"Graymaroon Cover" - 1936. Author's collection

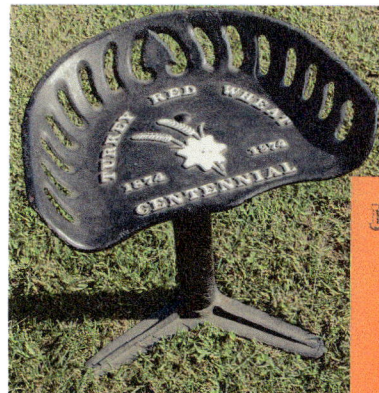

Implement seat designed and cast by Virgil Litke

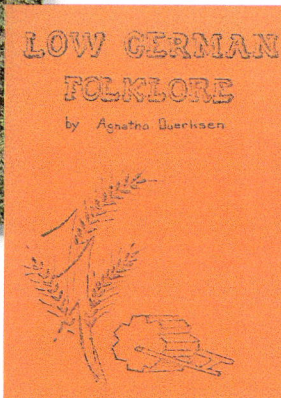

"Low German Folklore" booklet by Agnetha Duerksen. Author's collection

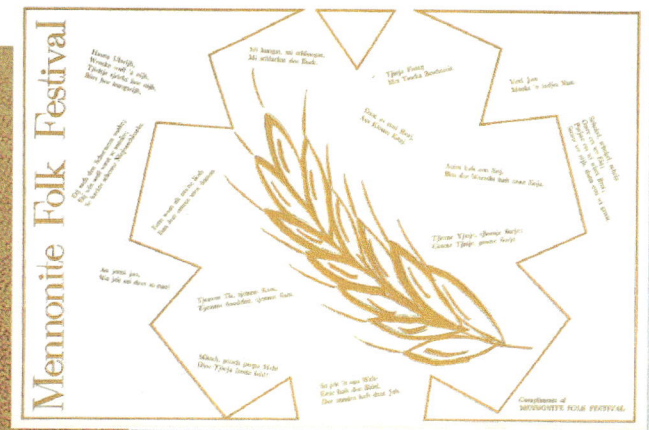

Folk Festival Poster - 1960. MLA

The "Bethel Collegian" - 1957. MLA

The "Bethel Collegian" - 1960. MLA

The "Bethel Collegian" - 1971. MLA

The "Bethel Collegian" - 1977. MLA

Proposed Fine Arts Center and Student Union both based on the threshing stone shape - 1961. MLA

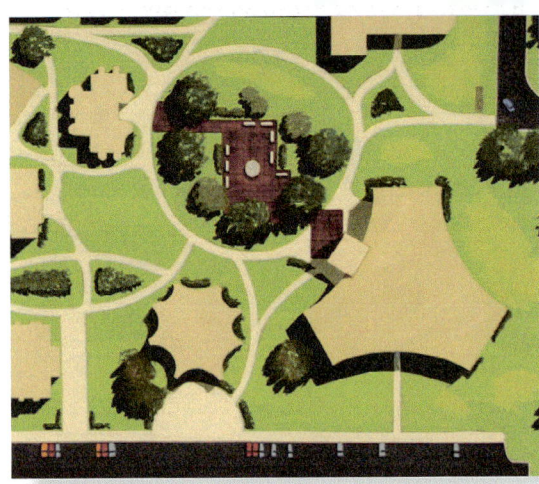

Architectural concept drawings for a Wheat Museum proposed by Harley Stucky and Kansas Wheat Commission - circa 1974. MLA

Bethel College Master Works Program - 2010. MLA

Model made of plaster and wood by Tabor College in 1974. Author's collection, gift from Ethel Abrahams

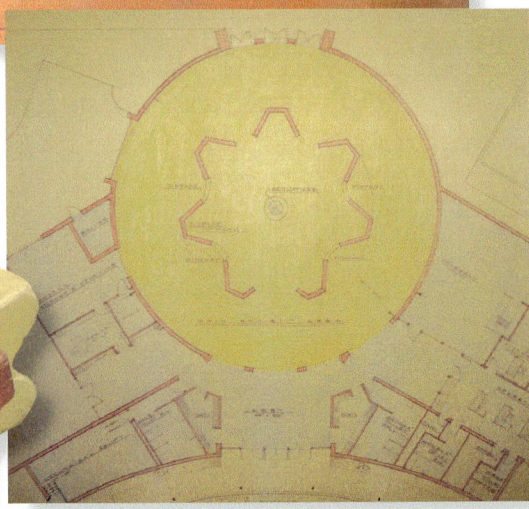

Threshing Stone #95

Most all my friends are probably tired of my constant talk and questions about threshing stones, but sometimes it pays off.

Tim and Judy Goertzen were at a gathering of friends and said that they had been talking to Judy's parents from Topeka, Kansas, and they thought that a fellow church member might have a thresh-

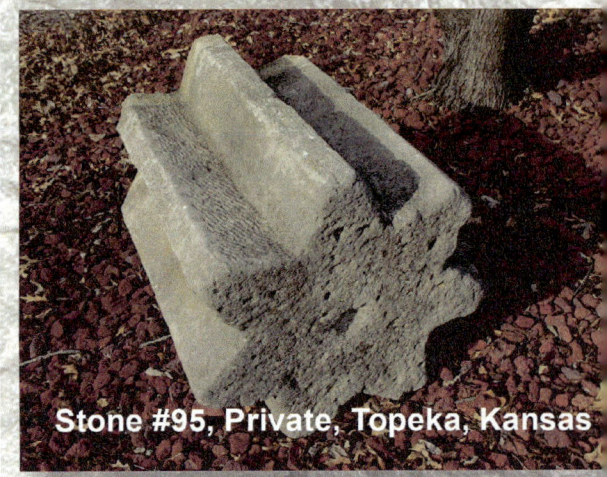

Stone #95, Private, Topeka, Kansas

ing stone. I got their name and made arrangements and we meet them on a Sunday in Topeka at the Southern Hills Mennonite Church. We met Phil and Patricia Esau and attended church, then had lunch followed by a photograph session at their home.

This stone comes from the Inman, Kansas, community from the family farm south and west of town.

This stone has been the focal point of many pictures over the years at several locations. The first two, black and white photos, are taken at the Elias Esau farm in the late 20s, with the women in one photo, and the men in the other, the eldest two being John Esau and his wife Maria. The next photo was taken at the John H. Esau farm in 1986.

"Consider the Threshing Stone" by Jacob J Rempel. Author's collection

Kansas Wheat Centennial Highway 50 road sign in Florence, Kansas

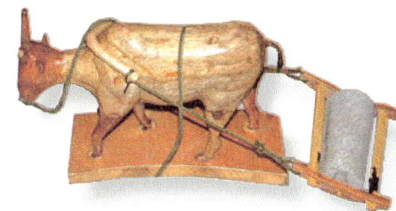

Chinese Jiangnan Historical Miniature Farm Implement carving. http://www.jxcn.cn/34/2009-10-19/30132@590751_1.htm

Kauffman Museum Sign - 1972. MLA

Bethel College Shirt. Author's Collection

Kauffman Museum sign - 2012

Food from Amazing Wheat

A grain of wheat is nothing short of amazing; it is the powerful grain that shaped human history. Civilizations have used wheat as a main source of food for longer than any other single food source. In fact bread may be the ancestor of all prepared foods starting nearly 12,000 years ago.

What makes this grain so powerful and important? Wheat is now grown on more land area than any other crop. It is third in total world production only behind rice and corn, and supplies about 20 percent of the food calories for the world's people where it is a national staple. About one third of the world's people depend on wheat for their nourishment.[1]

Wheat is the major ingredient in most breads, rolls, crackers, cookies, biscuits, cakes, doughnuts, muffins, pancakes, waffles, noodles, pie crusts, ice cream cones, macaroni, spaghetti, pizza, and many prepared hot and cold breakfast foods.[2]

Wheat is the history of civilization. Wheat allowed for the advancement of organized communities able to remain in one location, not being required to follow the hunt, and advancing the practice of organized trade from community to community, now with the ability to store food for the future.

Bread provided ancient people with a reliable food source which would keep through the winter months. This allowed them time to develop other useful skills beyond what was required to feed themselves.

It was not until the late 19th century that wheat cultivation flourished, including the importation of an especially hardy strain of wheat known as Turkey Red wheat to the United States. Russian immigrants who settled in Kansas brought Turkey Red wheat with them.[3]

One grain of domesticated wheat cannot reproduce itself without human intervention known as farming. However, when done correctly one grain of wheat can be prolific; a single kernel can grow several hundreds of seeds next year.

Each seed produces a clump of wheat, yielding several heads per clump, and each head yields up to about 50 kernels. Wheat is sown at a rate of about one bushel per acre and may yield about 50 bushels per acre today.

A bushel of wheat weighs in at about 60 pounds. A pound of wheat may contain 15,000 to 17,000 seeds, so a bushel may contain over 1,000,000 seeds.

We can process these wonderful seeds into an almost limitless resource, with many possible uses.

Opposite - Heads of ripe wheat, wheat and chaff, grains of wheat, flour, and Zwieback.

Food is by far the most valuable end product for the grains of wheat, feeding humans and animals with carbohydrates. Besides being a high carbohydrate food, wheat contains valuable protein, minerals, and vitamins. Wheat is also an efficient source of protein, when balanced by other foods that supply certain amino acids.[4]

Much of the wheat used for livestock and poultry feed is a by-product of the flour milling industry.

Bread is certainly the primary end product for wheat. Bread can be unleavened or leavened with yeast. When flour comes in contact with water and remains idle for a period of time, it begins to rise. In modern processes, yeast is added to aid in the rising, but even without yeast, dough will begin to ferment, and the resulting gases will cause the dough to rise. The Egyptians were the first to discover that this process would produce a light, expanded loaf. The Egyptians also invented a closed oven in which to bake the bread.[5]

The ancient Hebrews were in such a hurry to get away from their Egyptian captors that they made their bread

Homemade unleavened bread.
Wikimedia Commons - Matzo Yoninah

without leavening. Today Jewish people celebrate Passover, their escape from the Egyptians, with unleavened bread—"matzo." Bread without leavening also represents truth in Jewish tradition, because bread that is unleavened retains the true flavor of the grain from which it is made.[6]

Traditionally, people made bread from whatever grain grew best in the area where they lived. Wheat, rye, corn, barley, millet, kamut, and spelt are some of the grains used around the world. Wheat flour is preferred because of its gluten content. Gluten is what gives bread its elastic quality.[7] More gluten in a flour makes it easier for the flour to build up a tough structure able to trap the waste gases of yeast during kneading as well as to rise effectively during baking. Less gluten in a flour produces a lighter, less chewier texture such as those found in cakes.[8]

Bread is such a powerful food that ancient Egyptian governments controlled its production and distribution as a means of controlling the populace. In France the shortage of bread helped start the French Revolution.[9]

In the middle of the nineteenth century, a Swiss engineer invented a new type of mill with rollers made of steel which operated one above the other and

The stone was later moved to Topeka in 1991 where the last photo with Phil and his family was taken in 1998.

The stone is now located in front of their home, setting on its side in a flower bed. The stone is in excellent condition. It does not have an axle hole, but a dimple at each end where the axle would be.

Threshing Stone #96

This stone was actually on my radar for a very long time, but I finally got a break and found it. An antique collector in Newton told me of a guy west of Wellington that collected antiques and thought that he had a threshing stone, but he could not remember his name. He told me of another guy that would know his name but he did not either. I asked Jerry Toews if he knew of this guy and he thought he did but could not remember his name, but called me later and told me his name, but then I lost the slip of paper I had written it on. I called the first guy back a year and a half later on another lead and he remembered his name this time.

I called up Ed Larson who lives south of Milan, Kansas, and made arrangements to photograph it later that week. Ed has a wonderful collection of farm antiques and in particular a huge collection of threshing machines and old combines. It would only be fitting that he would have a threshing stone too.

Stone #86, Private, Milan Kansas

His stone, stored in a shed; is typical in size and shape; however it has the most fossil holes that I have seen in any stone.

He bought the stone from J. R. Janista at an auction, who had collected two stones that he believes originally came from the Goessel area. The second stone he believed had been

were driven by steam engines. Meanwhile, the North American prairies were found to be ideally suited to growing wheat. This, together with the invention of the roller-milling system, meant that for the first time in history, whiter flour, and, therefore whiter bread, could be produced at a price which brought it within the reach of everyone—not just the rich.[10]

One bushel of wheat yields approximately 42 pounds of white flour or 60 pounds of whole wheat flour. A bushel of wheat yields 42 one-and-a-half pound commercial loaves of white bread or about 90 one-pound loaves of whole wheat bread.[11] There are approximately 16 ounces of flour in a one-and-a-half pound loaf of bread.

Both whole wheat flour and all-purpose (white) flour are made from kernels of wheat. A wheat kernel is divided into three major parts—bran, endosperm, and germ.

All-purpose flour is made from only ground endosperm. Whole wheat flour is made by grinding the entire wheat kernel. When wheat is made into white flour, the outer layer of the kernel is separated from the rest of the wheat. The wheat bran and germ that have been removed are used in animal feeds.[12]

Flour:
There are several kinds of wheat flour available for sale with the most popular being enriched and bleached all-purpose flour. The differences between the flours comes down to the type of wheat, the parts of wheat

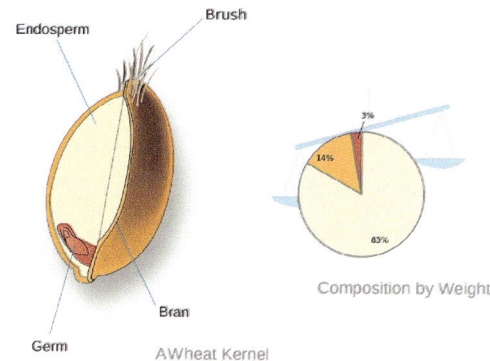

	Carb./g	Protein/g	Fat/g	Fiber/g	Iron (% daily req.)	Others
Bran	63	16	3	43	59	vitamin Bs
Endosperm	79	7	0	4	7	
Germ	52	23	10	14	35	vitamin Bs omega-3/6 lipids

Nutritional Value (per 100g)

A wheat kernel and its nutritional value.
Wikimedia Commons - Berghoff -

included, the processing of the wheat, and any additives added to the flour.[13]

All-purpose Flour:
High gluten flour and bread flour are produced from hard wheat. High gluten flour has a gluten percentage of about 12-14% while bread flour contains about 10-13% gluten. Both flours are almost completely made of hard wheat, but some high gluten flours are treated to reduce starch content, raising the gluten content to around 14%. These flours are generally used for making breads. High gluten flour is reserved for breads that are extra elastic such as bagels and pizza.[14]

Bulgur:
The oldest recorded use of wheat is in the form of bulgur. Bulgur is made by soaking and cooking the whole wheat kernel, drying it, and then removing part of the bran and cracking the remaining kernel into small pieces. Its uses are numerous, from salads to soups, from breads to desserts. It is a nutritious extender and thickener for meat dishes and soups. Tabouli is a popular salad made with bulgur, parsley, cucumber and tomato.[15]

Cracked Wheat:
Cracked wheat is very similar in nutrition and texture to bulgur. It is the whole kernel broken into small pieces, but has not been pre-cooked and dried. Cracked wheat can be added to baked goods for a nutty flavor and crunchy texture. Only a small proportion of cracked wheat can be used in breads, because it is very sharp and will cut gluten strands.[20]

Wheat Germ:
The germ of the wheat kernel is often added to baked goods, casseroles and even beverages to improve the nutritional value and give a nutty, crunchy texture. The protein quality of wheat germ is very comparable to that of milk. One-fourth cup of wheat germ contains about 110 calories.[16]

Wheat Bran:
The bran is the outer layer of the wheat kernel, often used for animal feed. It also makes a nutritious addition to baked goods, because it is a good source of fiber and is high in B vitamins, protein and iron.[17]

Wheat Berry:
The wheat berry is another name for the wheat kernel. The cooked whole kernel can be used as a meat extender, breakfast cereal or as a substitute for beans in chili, salad, and baked dishes.[18]

Cereal:
Many commercial cereals on the market are made from wheat and can be eaten as a snack, breakfast cereal, or added to baked products. A variety of ready-to-eat wheat cereals are available. The wheat may be shredded, puffed, flaked or rolled. The bran may be in the form of flakes or granules.[19]

donated to the Caldwell History Museum. I had a great time looking at his collections and learning considerable information about early threshing equipment. He had also made his own concrete form and had Yoders Ornamental Concrete cast him an actual sized reproduction threshing stone, which he had displayed in front of his house.

Threshing Stone #97
Leaving Ed's farm I headed south to Caldwell, where Ed believed the stone was displayed in front of the Museum at the main intersection in town.

As soon as I got to the center of town, there it was. A nice, typical stone displayed in a flower bed at the intersection of Old 81 and Main Street.

Stone #97, Caldwell Kansas Museum

Threshing Stone #98

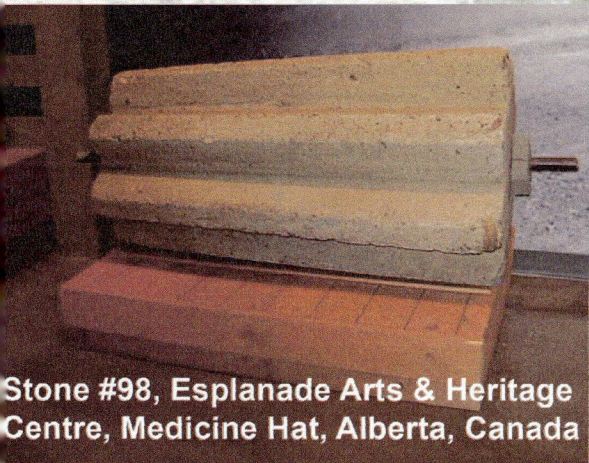

Stone #98, Esplanade Arts & Heritage Centre, Medicine Hat, Alberta, Canada

Photo courtesy Mel Bender

I received an email from Mel Bender, who read an article in the Summer 2012 AHSGR Newsletter about MY threshing stone project, and sent me this picture.

The threshing stone is in the Museum - Esplanade Arts & Heritage Centre, Medicine Hat, Alberta, Canada. A plaque next to the stone says, "Thresher, ca. 1905: This unique thresher was used to separate grain from its stalk. It was made by Samuel Wutzke, one of the thousands of Germans from Russia who homesteaded in the Medicine Hat district."

Wheat Flakes: Whole-wheat kernels that are steamed, flattened through rollers, and flaked, retain most of their nutrition, as they become a form of uncooked grain, not a ready-to-eat cereal. They can be used in many recipes similar to oat flakes.

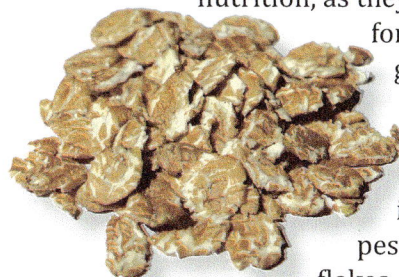

Non-food:

Non-food wheat products include wallboard, cosmetics, pet foods, newsprint, paperboard, and other paper products, soap, trash bags, concrete, paste, alcohol, oil, and gluten.[21]

Wheat is even used as the basic unit of weight in the British system. It was originally the weight equivalent to that of a single grain of wheat known as a "Grain". The grain is the smallest unit of weight in the Troy and Avoirdupois systems. The Avoirdupois pound is equivalent to 7000 grains, and then each pound is split into 16 ounces. Grains are currently used in the United States and Canada to measuring the mass of bullets and gunpowder.[22]

Let's take a look at some of the traditional foods made with wheat that are part of the Mennonite tradition.

100-year-old Zwieback. Author's wife's grandparent's wedding from 1910.

Zwieback:

Probably the most common form of Mennonite bread would come in the form of a *Zwieback* (also spelled Zwiebach or in Low German Tweeback). The Russian Mennonite Zwieback is a yeast bread roll formed from two pieces of dough that are pulled apart when eaten. Placing the two balls of dough one on top of the other so that the top one does not fall off during the baking process is part of the art and challenge that must be mastered by the baker.

Russian Mennonite Zwieback originated in the port cities of the Netherlands or Danzig, where toasted, dried buns were used to provision ships. Mennonite immigrants continued this practice centuries later.[23]

My mother typically would bake Zwieback nearly every Saturday morning. It was a staple food in our home and in many Low German Mennonite homes.

The same dough recipe would also be used to make Plautz (almost like a fruit pizza) where the dough would be pressed out in a shallow rectangular pan with seasonal fruit like cherries, mulberries, or peaches laid on top of the Zwieback dough then sprinkled with sugar and baked. Plautz is a wonderful summer desert, often enjoyed by my family for faspa in the field.

Zwieback (Tweeback in Low German).

Here is a Zweback recipe in the book *From Pluma Moos to Pie* compiled by Ruth Unruh and Jan Schmidt:[24]

--4 cups milk (or 1 cup dry milk dissolved in 4 cups warm water)
--1 1/2 cups shortening (butter and lard/shortening)
--1/2 cup sugar
--1 teaspoon salt
--3 tablespoons dry yeast, dissolved in 1 cup warm water
--8-10 cups all-purpose unbleached flour

Mix ingredients in mixer with batter beater. Add enough flour until dough is difficult to mix. Let dough stand 10 minutes. Then add more flour, a little at a time, using dough hook or kneading with hands to form a soft dough. Let rise and knead down once. Let rise again and shape into zwieback by pinching off balls of dough about 1 1/4 to 1 1/2 inches in diameter for the base and slightly smaller for the tops. Ruth Unruh would fill one cookie sheet with bases, flattening them slightly, and then another cookie sheet with tops. At that point, when they had set a couple minutes, she would put a top bun on each of the base buns, using the second knuckle of her forefinger or middle finger to press/pinch the top bun quite firmly down to the bottom bun. Let rise 5 minutes and bake at 400 degrees for about 10 minutes or until golden brown.

Bread:

Of course bread, particularly white bread, was the star of the Kansas hard winter wheat. The high gluten and white-white flour was the desired flour throughout the world. Bread is a staple food prepared by cooking a dough of flour and water and additional ingredients. Fresh bread is prized for its taste, aroma, quality, appearance and texture.

Mennonites have a long history of making breads according to Norma Jost Voth in her book *Mennonite Foods & Folkways From South Russia*. Busy Russian Mennonite housewives baked bread at home. Bakeries really did not make a difference in their lives. These pioneer women lived too far from towns. What is more, they were frugal. It was more economical to bake at home. To spend even 5 cents for a loaf of bread did not seem right.[25]

Whole Wheat Bread:

Here is a recipe for whole wheat bread from my mother-in-law, Ruth (Schmidt) Unruh of Goessel, Kansas, a frequent Kansas State Fair blue ribbon winner for her breads:
--4 tablespoons yeast
--1-1/2 cups potato water (unsalted)
--Dissolve yeast in potato water, set aside
--2-1/2 tablespoons salt
--2 tablespoons sugar
--7 cups all-purpose (white) flour

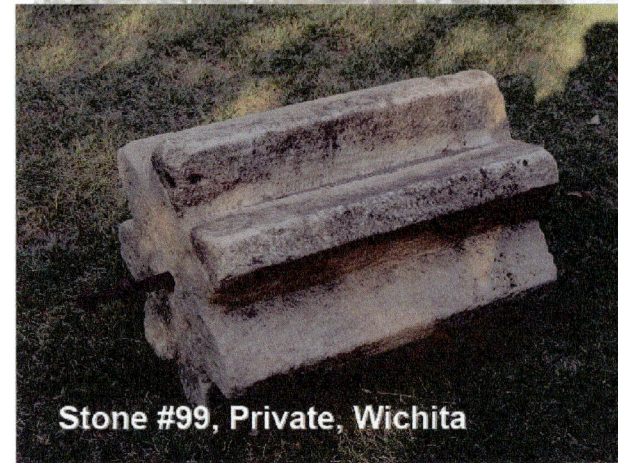

Stone #99, Private, Wichita

Threshing Stone #99

This stone also came to my attention from both Ed and the antique collector in Newton.

The antique collector had told me a year and a half earlier that he knew where another threshing stone was but he was reluctant to tell me because the owner was elderly and he did not feel comfortable giving her name out.

So a year and a half later I asked him again, and he said she had since passed away and he told me who the son was. That same day I met with Ed Larson, he too knew of this stone and gave me the name of the son and told him that I would be calling.

The stone is now in the possession of Hal Ottaway. His

father Harold Ottaway had been an avid antique collector and had given the stone to him. It was nicely displayed in the beautiful yard at his residence in Wichita.

Threshing Stone #100

The keen eye of Ann Birney (See stone story #89) spotted this stone in Lipscomb, Texas. Ann and Joyce Thierer had been at the Wolf Creek Heritage Museum and were quite sure they had seen a threshing stone in a row of farm implements.

I contacted the director Virginia Scott, and she sent me the image of the unique threshing stone with 10 teeth.

The town is about 60 miles south of Liberal, Kansas, and is home to some Mennonite residents, but the provenance is still unknown.

Stone #100, Wolf Creek Museum Lipscomb, Texas

Baked Bread, a staple around the world.

Add salt and sugar to 1-1/2 cups potato water. Beat in 3 cups flour. Add yeast.

Then add 3-4 cups flour and mix with spoon until elastic. Cover and let rest 15 minutes.
--1/2 cup hot water
--1/2 cup + 2 tablespoons brown sugar
--4 tablespoons shortening
Combine hot water, brown sugar, and shortening, cool to lukewarm. Then add to yeast mixture.
--6 cups whole wheat flour
--All-purpose (white) flour as needed
Add whole wheat flour and knead. Add all-purpose flour as needed to make a soft dough. Knead until smooth and shiny (about 8-10 minutes).

Place in greased bowl, turning to grease top. Cover; let rise in warm place, free from draft, until double in bulk, about 1 hour.

Punch down. Divide into four balls, shape into loaves. Place in four greased 9 x 5 x 3 inch bread pans. Cover; let rise till double in bulk, about 1 hour. Bake in preheated oven at 400 degrees F, 25-30 minutes, or until hollow sounding when tapped. Remove from pans immediately and cool on wire rack. Brush tops lightly with butter.

Verenike (Low German).

Verenike:

Verenike is a traditional Low German Mennonite ethnic food, a type of dumpling with cottage cheese filling and often covered in ham gravy or syrup. In Norma Jost Voth's book, *Mennonite Foods & Folkways from South Russia*, Diana Jost provided this recipe. It may be doubled or tripled.[26]
--2 cups all-purpose flour
--1/2 teaspoon baking powder
--1 teaspoon salt
--2 eggs
--1 cup heavy whipping cream

Verenike - Cottage Cheese Filling:

--Mix together:
--2 1/2 cups dry curd cottage cheese
--1/4 teaspoon salt
--2 egg yolks

Verenike - Ham Gravy:

For Ham Gravy, fry diced ham in pan with a small amount of oil. (For Sausage Gravy, fry bulk sausage, drain off most, not all, of the grease.) Add 3-4 tablespoons flour, stir, and fry. Add milk or cream and boil until thickened. Serve over Verenike.

Combine flour, baking powder and salt and stir together. Beat eggs lightly. Combine with cream. Gradually work in the flour mixture. Allow dough to

stand, covered, for an hour or more before using. Roll dough out quite thin and cut into 3-inch circles. Place a heaping teaspoon of Cottage Cheese Filling on each circle. Fold over and pinch edges together tightly. Drop Verenikje into a large kettle of boiling, salted water. Boil 5-8 minutes. Drain in a colander and brown in a heavy skillet. Serve with gravy and country sausage.[27]

The Cream Gravy to cover the Verenike has many variations. The basic is melt 2-3 Tbsp. butter in frying pan and allow to cook until slightly browned. Add 1 cup of heavy sour cream (preferably farm-style) and a dash of salt and heat through. Fried ham or sausage drippings add a good flavor.

New Year's Cookies:

The origin of the New Year's Cookie or Portzilke is unclear, but according to Norma Jost Voth in her book, "*Mennonite Foods & Folkways from South Russia*, Mennonites gave their Russian neighbors Portzilke when they came and sang for them on New Year's Day. The "cookies" were a symbol of affluence and luxury and carried with them the wish for an abundant year.[28] They are made from a spongy yeast

New Year's Cookies. Photo Melissa Frantz

batter with raisins and dropped by spoonfuls into hot deep fat. The name means "Tumbling Over" since they turn over by themselves when the dough puffs up, tumbles over, and fries to a golden brown. The "cookies" (or raisin fritters) are then rolled in sugar or glazed.[29]

This recipe used by Karen's mother, Ruth Schmidt Unruh, is from the *Off the Mountain Lake Range* cookbook:[30]
--2 cups milk
--1/4 cup butter
--3 eggs
--1/4 cup sugar
--1 teaspoon salt
--2 cups raisins
--1 compress yeast cake
--4 to 5 cups flour

Scald the milk. Cool to lukewarm and add remaining ingredients. Mix. Let rise until double in bulk. Drop into deep fat by spoonfuls and fry until brown.

Bohne Beroggi:

Bohne Beroggi is a Swiss Mennonite ethnic food. Along with the Low German Mennonites, the Swiss Russian Mennonites also settled into the central Kansas plains and brought with them some of their own food traditions. Bohne Beroggi is a dough pocket filled with pinto beans and covered in a sweet white sauce. It is often eaten as a dessert.

This recipe is the one used for making Bohne Beroggi at the Kansas Mennonite Relief Sale:
--2 cups pinto beans
--1-1/3 cups sugar

Yet to be found?

I have followed many threshing stone leads that have ended up at dead ends. I have traveled across the state, sometimes only to find that it was actually a mill stone. Many other times following leads ended with a threshing stone I already knew about. Some leads were just so vague there was little to follow. But all were fascinating to track and it was always a good day when I found a new stone.

I still have several strong leads to track, but time has run out for now. For every threshing stone I've found, there may be another yet to find.

The most elusive stone that I have yet to find has such a long story, I cannot tell it all here. As of now the trail ends with a slide that proves it was in a Smithsonian exhibit in Washington, DC in 1985. But where is it now?

Assuredly, as soon as I publish I will find more, so check www.threshingstone.com for updates.

The absolute number of threshing stones made or found

doesn't really matter, but the hunt was fun, and I think the results were interesting.

We can now conclude that not thousands were made, but just a few hundred. We also now know that this is mostly a 4-county Kansas phenomenon, and not spread out across North America. We can say that it is centered among Mennonite farmers. We can say that it was a short-lived phenomenon.

Like I have often said, "Leave no stone unturned," and if you are lucky enough to have one of these rare artifacts, cherish it, pass it on to the next generation, and tell the story.

Russian stones found?

The story continues in the next Article.

--1-1/2 teaspoons salt
--Your favorite sweet dough recipe using 2 cups liquid.

Wash and cook beans until tender. Pour off liquid. There should be no juice left on the beans. Mash beans and add sugar and salt. Make small buns from sweet dough and let rise 10 minutes. Pull apart and place a heaping teaspoon of the bean mixture in the center of each bun. Fold over and pinch edges together. Place on greased pans and let rise 10 minutes. Bake 20 minutes at 350 degrees.

Sauce for Bohne Beroggi:

--2 cups milk
--2 cups cream
--2-1/2 tablespoons cornstarch
--1 cup sugar

Heat milk and cream, add cornstarch and sugar mixed thoroughly, and cook until thick, stirring constantly.

Poppy Seed Roll:

Use your favorite sweet roll recipe or this recipe from Mae (Mrs. Paul) Schrag.
Sweet dough:
--Mix together, heat, and cool:
--2 cups water
--½ cup lard
--½ cup sugar
--2 teaspoons salt
Stir together in a separate

Bohna Beroggi (Swiss Mennonite bean dumpling).

bowl or glass measuring cup:
--4 teaspoons yeast
--1 scant teaspoon sugar
--½ cup warm water
--5 ½ to 6 cups flour

Poppyseed Filling:

--1 lb or 3 cups poppyseed, cleaned and finely ground
--3 cups sugar
--2 cups half & half

Add yeast mixture to cooled mixture with 5 ½ to 6 cups flour and mix together to form a firm dough. Let rise 45 minutes, punch down.

Prepare Poppyseed Filling while waiting for dough to rise. Bring to a boil and cool before spreading on dough. Grease and flour two 9 x 13 pans.

After punching down dough, let rise again for 45 minutes. Punch down and form into 6 large buns about the size of a fist. Let rise again until double in size. Punch down and roll out one of the buns very thin, about 3/16 inch thick and about 12-13 inches long on one side. Spread 1/6 of cooled Poppyseed Filling over the rolled out dough. Roll up dough like a jellyroll, seal the edges, and place lengthwise in greased and floured 9 x 13 pan. Repeat with each of the remaining 5 buns. Three loaves will fit side by side in a 9 x 13 pan.

Bake in 325-degree oven for 25-35 minutes until golden brown. If desired, brush with beaten egg before baking.

Poppy Seed Roll (Mach-Kuchen, Swiss Mennonite).

Google™
SEARCH WEB
SEARCH IMAGES
TRANSLATE
MAPS AND STREET VIEW

Mennonite Library and Archives

BETHEL COLLEGE
-CONTEXT
-THRESHER E-VIEW

Wheat Heritage Engine and Threshing Co. Inc.
BOOTH

facebook
THRESHINGSTONE PAGE

www.threshingstone.com
CREATE A WEBSITE

MENNONITE **HERITAGE** AND **AGRICULTURAL** MUSEUM

Global Anabaptist Mennonite Encyclopedia Online

HOW TO FIND THRESHING STONES

American Historical Society of Germans from Russia
RESEARCH

MARION COUNTY MUSEUM

craigslist

Babelfish Translation

The **Kansan** .com
ARTICLE

WORD OF MOUTH

RENO COUNTY HISTORICAL SOCIETY

Kansas Museums Associaion
NEWSLETTER

HARVEY COUNTY HISTORICAL SOCIETY

BETHEL COLLEGE FALL FEST
BOOTH

MPMA
NEWSLETTER

INMAN SANTA FE TRAIL DAYS
BOOTH

GOESSEL THRESHING DAYS
BOOTH

GERMANS FROM RUSSIA HERITAGE COLLECTION

ebaY
SEARCHES

KANSAS HISTORICAL SOCIETY
RESEARCH

KAUFFMAN MUSEUM

BUHLER FROLIC
BOOTH

CENTER FOR MB STUDIES
RESEARCH

Mennonite Weekly Review
ARTICLE

Kansas Sampler FOUNDATION
NEWSLETTER

McPHERSON COUNTY HISTORICAL SOCIETY

MENNONITE SETTLEMENT MUSEUM

Finding the Stones

There is no book or YouTube video that tells you how to find threshing stones. You have to make that up as you go. Who knew that the hunt would be so much fun. I had no idea I would enjoy the "treasure hunt" so much. Finding lost treasures has been an epic and rewarding process. Here is how I did it.

Finding leads, then tracking those leads created very circuitous routes of discovery. Most of all I enjoyed talking with individuals to learn more about their stones' history. My motto became "Leave No Stone Unturned" as I would track every single lead to ultimately finding the stone or disappointingly having the lead dry up and fade away.

I had no idea how to start my quest. I of course knew of the one in our front yard that started this whole project...I knew Bethel had a few...I knew of at least one at Kauffman Museum, and I knew I had seen some at Goessel's Mennonite Heritage Museum. But how many were there and how would I find them?

I think I grew up knowing of threshing stones. After all, I had gone to Bethel College athletic events my whole life, so I knew what a threshing stone was because the mascot of Bethel is "The Threshers." But actually I did not recall any from my home community around Buhler, Inman, and Hoffnungsau Mennonite church. So how many are there really?

My naive assumption was that there are thousands of threshing stones spread out across the country wherever wheat had been grown. But to find out I started by asking people who I thought might know, but got few answers. Brian Stucky, a friend of mine and a local historian, had discussed threshing stones with **Assumption** me several times over the years and he thought that there were very few, maybe only a hundred or so. Wow, I was shocked. Not thousands? He said he had started a list once and came up with maybe 15-20 stones, but he did not **I Need a Plan** remember exactly how many and did not recall where his list may have ended up. So I still needed a plan.

Obviously word of mouth is the best way to start, so a few names quickly emerged as the people to talk to. John Thiesen, Co-Director of the Bethel College Libraries and Archivist had given a

Opposite page: Resources used to find the threshing stones.

faculty lecture titled "What's a Thresher? Recovering Bethel's Symbolic History," so I talked to him. Jerry Toews, a local antique collector and a leading expert on steam-era equipment was soon mentioned, so I contacted him. Also I contacted the local history experts, Rachel Pannabecker at Kauffman Museum, Ray Wiebe,

Who are the Experts?

a retired historian and author, and an old friend of mine Kelly Harms who is an avid tool collector and is knowledgeable about local farm history.

Well with a few names and a list of objectives I needed to get started. What was my plan and how did I track my finds? I started with a simple notebook, by writing down what I knew, and added to it as I found out more information. I decided to develop a spreadsheet, to log my discoveries, starting with num-

A Log

bering the stones in the order that I found them, ours being number one. I felt it best to keep this list just to myself, in order to keep the privately owned threshing stones confidential; only I and

the owner would know for sure.

This privacy would also allow individuals to be more willing to share their information. This, I think, was a good decision. But even so, a few persons were still reluctant to give out much information.

Privacy

The spreadsheet includes the name of the current owner, the address, the setting in which the stone is located, special observations about the stone, such as the dimensions and number of teeth, the GPS location, and most importantly a picture.

I talked to several individuals who had interest in threshing stones in the past but did not document their discoveries, so I decided early in the project to document as much as I could and that a photograph would insure that my research was credible. Telling the stories of finding the

Proof

stones without divulging names and locations is somewhat difficult but I will you tell what I can. In some cases I have obtained permission to use the names of the people who helped me find the stone and the names of the people who have owned the stone through its history.

Threshing stones found in Russia

Thanks to the many individuals that have been in contact with me and given permission to use the following photos.

Threshing stone bench
Neukirch, Ukraine.
1999
Photo - John Esau

Mennonite home with two stones
Rudnerweide, Ukraine.
1999
Photo - John Esau

Rudnerweid, Ukraine.
1999
Photo - John Esau

Near Alexanderwohl, Molotschna, Ukraine.
Photo - Arlin Buller

Concrete threshing stone.
Photo - Stan Hill

Gnadenfeld, Ukraine Agriculture Museum, 30-40 stones.
Photo - Stan Hill

Gnadenfeld, Ukraine Agriculture Museum, 30-40 stones.
Photo - Stan Hill

Gnadenfeld, Ukraine Agriculture Museum, 30-40 stones.
Photo - Stan Hill

Gnadenfeld, Ukraine Agriculture Museum, 30-40 stones.
Photo - Stan Hill

The process has been fun because I never knew what I might find. There is no list of stones that were made, they are not serialized with sequential manufacturing numbers, there are no patents, no list of manufacturers, just an unknown number over an unknown area. The quest was on.

What I found surprising was how little people knew about these stones. This led me to believe that maybe an article or even a book might be appropriate to tell the story.

Like all history projects, starting the project about 100 years earlier would have been very helpful, allowing me talk to the actual users of threshing stones and getting the facts directly from them, but that didn't happen until nearly the end of my research. (See Article 11, interview with George Becker.)

One is left only with legacy stories from the people who are very elderly. As I found, they may also be the hardest to communicate with, either through loss of hearing, loss of memory, or just a bit of confusion.

But I made it a point that if any person was willing to tell me about their history, I would spend all the time with them that they wanted. I learned quickly that people love to talk about their families and memories. The longer people talked, the more they remembered. Whether young or old, talking alone brings back memories.

So many times I was glad I did not cut the conversation short, because additional information would finally come out. With some persons, several conversations were held, and then I also learned that stories may change the next time I heard them. The other challenge was that if I interviewed several persons on the same stone, there might be different memories that each recalls. But I suppose that I discovered for myself that this is the challenge of those that write history.

Big Project!

It was never clear to me how big this project would become, how much interest there would be in it, or how much time it would take. But the process was fun. I started with what I knew and then kept adding to my knowledge base.

How to get the word out?

Word of mouth was by far my best ally, but I also tried every technique I could think of to get the word out to the public. Here are some methods I used.

I received great help from Bethel College. When they heard about my project they soon realized that this could tie in nicely to the Bethel 125th Anniversary Celebration. This tied me in to great public relations opportunities. Melanie Zercher wrote an article and got it pushed out to the public through the "Thresher

Poster used at festival table displays.

Do You Know Where a THRESHING STONE is located?

History Research Project

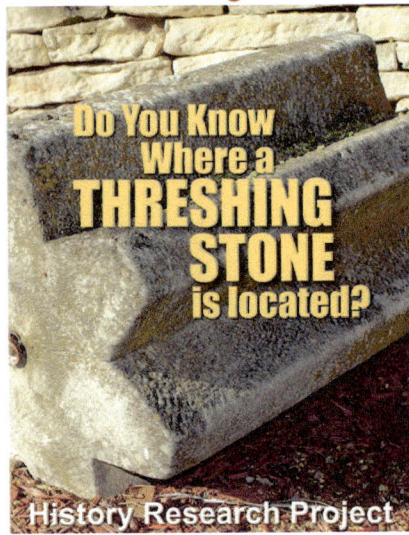

E-View," as well as local papers such as "The-Newton Kan-san," "Hillsboro Star-Journal," "Marion County Record," "Buy-ers Edge," "The Ledger," "Hutchinson News," and "Hillsboro Free Press."

Regional stories went out through Kauffman Museum Newsletter and the AHSGR Newsletter. Additional stories were posted via the internet by Kansas Sampler Foundation, Mountain-Plains Museums Association, Kansas Museums Association, and Mennonite Historical Society of Saskatchewan.

Use the Press

International press came from a story that was published in the "Men-nonite World Review." I also got broad press cover-age when I received the Bethel College 2012 Outstanding Alumni award where I made sure the press release included

information about my project. This article was reprinted in many other newspapers.

I also used the resources of all the following institutions for help to find threshing stones or for infor-mation about threshing stones: Kauffman Museum, Mennonite Library and Archives (MLA), Center for Mennonite Brethren Studies, Goessel Mennonite Heritage and Agriculture Musuem, Kansas State His-torical Society, Harvey County Historical Society, Reno County Historical Society, Marion County Historical Society, McPher-son Museum and Arts Association, Harvey House Museum, Smithsonian Museum, American Historical Society of Germans from Russia (AHSGR), and Germans from Russia Heri-tage Association.

I did face-to-Face interviews with many individuals by attending many com-munity celebrations such as: Goessel Threshing Days, Buhler Frolic, Inman Santa Fe Days, and Bethel College Fall Festival. To get attention at these events I made posters and t-shirts.

Institutional Resources

Local Festivals

LEAVE NO STONE UNTURNED
www.threshingstone.com

Ukraine stone.
Photo - Stan Hill

Ukraine stone.
Photo - Stan Hill

Gnadenfeld, Ukraine Agriculture Museum 30-40 stones.
2010
Photo - Becky and Don Linscheid

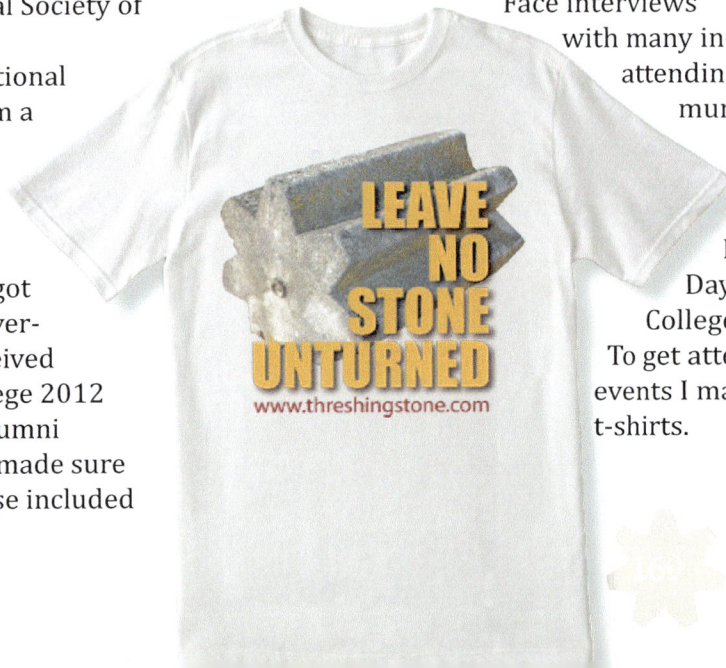

Gnadenfeld, Ukraine Agriculture Museum 30-40 stones.
2010
Photo - Becky and Don Linscheid

Gnadenfeld, Ukraine Agriculture Museum 30-40 stones.
2010
Photo - Becky and Don Linscheid

Gnadenfeld, Ukraine
Agriculture Museum
30-40 stones.
2010
Photo - Becky and
Don Linscheid

Gnadenfeld, Ukraine
Agriculture Museum
30-40 stones.
2010
Photo - Becky and
Don Linscheid

Neukirch, Ukraine
Memorial to 7 vil-
lages.
2008
Photo -Ralph
Kauffman

Neukirch, Ukraine
Memorial to 7 vil-
lages.
2008
Photo -Ralph
Kauffman

The above memorial used threshing stones re-
covered from the surrounding former Mennonite
villages. A local high school history teacher taught
his students about the history of the Mennonites.
The memorial was built by his high school students
to commemorate the people of 7 villages who suf-
fered from famine, holocaust, and deportation. There
are no Mennonites or Germans in that region since
WWII. A total of 8 threshing stones.

Internet Resources

I used the internet as an endless resource, both for finding stones and for research; using Google search, Google maps, Google street view, Google translate, Babelfish translate, and Yahoo translate. I could work in many languages and this allowed for website and image searches around the world. I also was able to communicate with individuals in Russia that could not speak English, amazing!

I also had to learn how to store all my data, photographs, interviews and notes on many books. I decided to do it all digitally. Fun but a whole new challenge.

Learning how to buy, build and manage a web-site was another education for me. I am not a geek so I really had to learn from scratch. Please check it out sometime: www.threshingstone.com. The end result was positive and it has seen a lot of world-wide traffic. I also set up a Facebook page "Threshing Stone." I hope

My Web Page

to update these more regularly in the future so people can keep up on things I learn after publishing this book.

The best part of this project was connecting with many individuals, such as the Richards family of Emporia and Kansas City. Finding their stone created a family reunion of sorts and gave my wife Karen and me a chance to learn to know yet another wonderful family in the process of doing research.

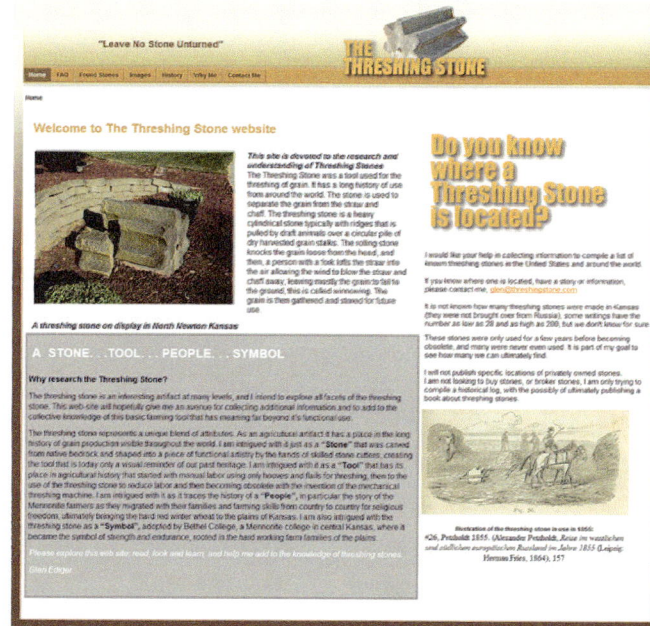

I used every resource I could thing of, and that is part of what made this project so much fun - "connecting the dots" as they say.

The family associated with "The Story of Kansas" history books. Barry Felkner, Milli (Richards) Felkner, Ann Birney, and Betty (Richards) Dunhaupt.

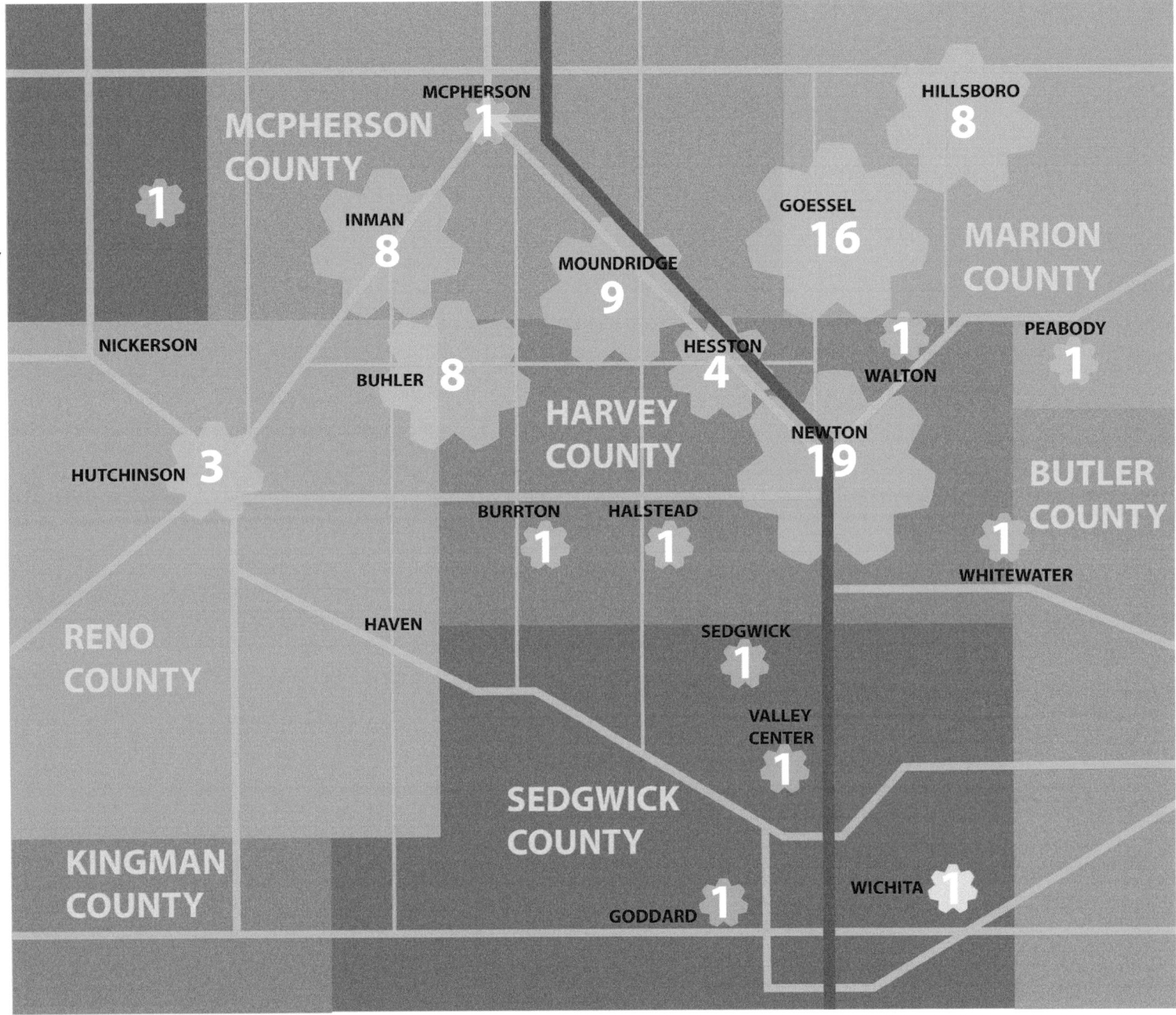

BISMARK NORTH DAKOTA 1

STEINBACH MANITOBA 2

MEDICINE HAT ALBERTA 1

LINCOLN NEBRASKA 1

TOPEKA KANSAS 2

KANSAS CITY MISSOURI 1

ELKHART INDIANA 1

CANYON CITY COLORADO 1

MCPHERSON COUNTY

MCPHERSON 1

1

INMAN 8

MOUNDRIDGE 9

GOESSEL 16

HILLSBORO 8

MARION COUNTY

NICKERSON

BUHLER 8

HESSTON 4

WALTON 1

PEABODY 1

HARVEY COUNTY

NEWTON 19

HUTCHINSON 3

BURRTON 1

HALSTEAD 1

BUTLER COUNTY

WHITEWATER 1

HAVEN

RENO COUNTY

SEDGWICK 1

VALLEY CENTER 1

SEDGWICK COUNTY

KINGMAN COUNTY

GODDARD 1

WICHITA 1

ELK CITY KANSAS 1

LIPSCOMB TEXAS 1

MILAN KANSAS 1

CALDWELL KANSAS 1

Now What Do We Know

I love puzzles. This whole project is nothing but a big puzzle. Finding the pieces and putting them together has been nothing short of fun for me. I didn't find all the pieces but I found enough to form a mosaic of information complete enough to help us form a historical picture.

To be sure the use of the threshing stone in North America is a relatively small phenomenon, both in location and duration. I was personally very surprised to learn how regional this story is and particularly how briefly the stones were used.

This whole book tells the details learned from almost 3 years of research. What conclusions can be drawn from the information collected?

History can be our guide to begin with. As I have stated earlier if not for a multitude of world events happening at a given time we would not have any threshing stone to be found. While we know that threshing with rolling stones has been used since Roman times, the Mennonites and other

farmers appear to have started using them in the Ukraine from about 1840 until they became obsolete, being replaced by mechanized threshing machines.

1840-1900 in Ukraine

But records also tell us that during the Russian Revolution in 1919, these old stones were drug out from abandonment and used to thresh grain once again. A preponderance of photographic evidence shows that they were plentiful in that region and many can still be found today, either in museums, for landscape decoration, or just left in a field to be all but forgotten.

Since Roman Times

Opposite page: Central Kansas map of communities where the threshing stones were found, not necessarily their origin. (Map is intentionally kept vague to protect private ownership locations.)

100+Found in Russia

Bench made with Threshing Stones in Ukraine village of Neukirch in Molotschna. Photo John Esau

I have collected images of over 100 different threshing stones that remain to this day in Ukraine. I have included images of some in the center column.

As you can see these are very similar in design to what we see in America. Almost all have 7 ridges, but the sizes vary a bit more, many longer than 30" and smaller than 24" diameter. But all have the typical axle hole with a steel shaft. It appears that the axle is firmly attached to the stone either by being tightly

Russian Concrete with end bump - Ukraine. Michael Clark

wedged in place or in some cases with steel straps that lash the axle to the stone.

Another common design difference seen in most of the Russian stones is a raised bump on each end which functions as a spacer to keep the ridges of the stone from contacting the towing frame. These appear to be a later design made of concrete.

Old Russian threshing stone drawing. http://forum.wolga-deutsche.net/viewtopic.php?f=4&t=894

Only a few images from Russia show the design of the towing frame, but as in this country all the original wooden frames have long rotted away. However a few steel frames still exist that show several styles that were created, the scissors style and the U-frame style. It appears that there are several colors of stone used in Russia varying from the common limestone color to a darker red-orange color seen on several stones.

Photos and drawing from many other countries throughout history can

Odessa Oblast, Ukraine - Museum Director: Nikolay Photo - Herb Poppke. 2003 Public Domain

Threshing stones 50 or more in a row Photo - Herb Poppke. 2003 Public Domain

Concrete threshing stone - Chortitza, Ukraine - Near the famous old oak tree. 2007 - Photo Rachel (Unruh) Clark

Molotschna Colony Grossweide, Ukraine. 1997 Photo - Lowell Ratzlaff

This threshing stone lay between the street and a yard in Lindenau, Ukraine. 1997 Photo - Lowell Ratzlaff

Grossliebental, Ukraine.
2008
Photo - Mel Bender

The Pushkin Museum of Fine Arts, Moscow. Discovered during the expedition of the Volga River Floating University in 2007.

Threshing stone in Omsk State History Museum.
2011
Photo - Peter Wiebe

"The stone is for threshing grain - used by rural population of Omsk region in early 19-20 centuries. One of these things is in a museum in the city Isilkul. It was accidentally found in a field near the village of Ksenevka, Isilkulskom region, Omsk region. Item is a stone block, hexagonal in cross section, which is set on a metal rod. The length of the subject - 96 cm, diameter - 35.5 cm in the vicinity of the Mennonite settlements. Two more instances of stone rollers are in the museum of Poltava, Omsk region, by ethnicity, these products are Ukrainians". Text - Peter Wiebe

Mennonites Perfected

also be found confirming the widespread use of the threshing stone. However many consider the Mennonites of the Ukraine to have perfected the design.

Mediterranean Many

So as far as finding threshing stones in other countries, there is evidence that wherever grains have been grown, the threshing stone has been used. All across southern Europe, all around the Mediterranean Sea, across eastern Europe, across the Middle East, across India, China, and even in Australia. However, I have found no evidence of the threshing stone being used in South America, by Mennonites or any other ethnic group.

Canada Not Much

Only three threshing stones so far have been confirmed in Canada: two in the Steinbach Museum and one in the Museum in Medicine Hat. But I also found a picture in an old Arab cookbook from the 1930s that talks about using the stone in Saskatchewan. They had gotten it from a Mennonite family that was no longer

Steinbach Museum - Roland Sawatzky - 2011

using it. However even by checking with individuals in that community as well as the Mennonite Historical Society of Saskatchewan inquiring for me, no others have been located. But I suspect if there was one, there were more.

China Many

One of many images found of threshing stones in China - www.cddsb.com/art/2007/10/25/art_8137_281118_1.html

In China threshing stones are treasured artifacts, but the line between mill-stone and threshing stone can be blurred at times. Many stones found have either very many ridges, or no ridges, and they are often of different dimensions, some being very huge and some with considerable taper. There is also a farming device that is found in the Guangdong Museum of Cultural History

Author's photograph of a wooden threshing device in Guangdong Museum of Cultural History - 2011.

in southern China that I was sure was a wooden threshing stone. It had a similar towing frame for pulling behind a horse, and had 7 ridges.

I was directed to what appeared to be a threshing stone, used as focal point in a landscaping magazine in the Summer 2011 issue of "Country Gardens." After considerable research I finally contacted the owner of this unique stone. It was actually a threshing stone that was shipped to the United States from Spain by a landscape designer who collected rare old artifacts to be used in upscale landscape projects in Virginia. He told me that he had imported up to a dozen or so from Spain; he said they were plentiful. I have not included these stones in my tabulation of found stones.

Not T-Stones

I was also directed to what appeared to be a whole line of threshing stones along a road on the east coast. They looked a little like threshing stones, but they were not.

So the threshing stone has been known around the world for along time but what about the ones that I found in the United States? I eventually found 100 threshing stones in North America, but to my surprise they were not all across the U.S., but ended up being a very, local Kansas story. At the beginning of this article is a map that shows where the 100 were found. There is no surprise now that they are still in the communities from which they were used, basically in a four county area.

Total Found 100

I have found many threshing stones that have moved from community to community by people buying them at farm auctions or by inheritance. But many are still on the farm on which they were originally used.

Swiss Volhynian?

I find no evidence that the Swiss Mennonites (Moundridge or Pretty Prairie) used threshing stones on their farms. This was surprising to me. Maybe they used other methods and then quickly advanced to the use of threshing machines.

Nebraska?

Even though many of the Nebraska Mennonites immigrated at the same time, I find no evidence that the Nebraska Mennonites owned or used threshing stones on their farms. I was told by one Nebraskan that they may have arrived at the same time as the Kansas Mennonites but were more prone to adopting new American technologies.

South Dakota? Minnesota? Oklahoma?

Collection of stones Ukraine.
2008
Photo - Ben Stobbe

Collection of stones 2008 - when the museum was in Gnadenfeld. Photo - Ben Stobbe (Friends of the Mennonite Centre in Ukraine)

Collection of stones Ukraine.
2008
Photo - Ben Stobbe

Collection of stones Ukraine.
2008
Photo - Ben Stobbe

The Pushkin Museum of Fine Arts, Moscow. 2007

Threshing stone on the ground close to Neu-Balzer.
2009
Photo - Alexander Spack

Threshing stone on the outskirts of Kauz (Vershinka).
2009
Photo - Alexander Spack

Threshing stones in the vicinity of Rothammel.
May 2009
Photo - Alexander Spack

Krasna, Bessarabia. 2008
Photo - Merv Weiss

Other than finding one threshing stone in North Dakota I have found no threshing stones used by Mennonites who settled in other Great Plains states.

Whitewater?

I was surprised to find no evidence that the Whitewater, Kansas community ever used threshing stones on their farms. This group of Mennonites migrated a year or more later to central Kansas, and most likely by that time there were many more threshing machine options, so the stone was never made or used there.

Volga Germans?

Despite many rumors, I found little evidence that Volga German immigrants that migrated to Kansas used the threshing stone. I have found only a few stones that do not appear to have any direct connection to Mennonites, so I suspect that they were made and used by some of these immigrants. In studying the fabrication of these stones, they all appear to be the same dimensions, design, stone, and show the same fabrication techniques. So very possibly they too got them from the same quarries and stone mason.

Paraguay?

Many Mennonites immigrated to Paraguay in the 1920s. Though some very old individuals do remember threshing stones in the Ukraine, there is no evidence to suggest that the Paraguayan Mennonites ever used threshing stones in the new country.

The Final Data

Location in North America

Kansas	91
Nebraska	1
Colorado	1
Missouri	1
Indiana	1
North Dakota	1
Texas	1
Canada	3

Still on the Original Farm

Only about 10 are still on the farm on which they were originally used.

Full Size Replicas

Rod Funk made about 20
Galen Froese made some replicas
Yoders Ornamental Concrete

Number of Teeth

All in North America have 7 except for the one in North Dakota which has 5, and one in Texas with 10.

On Public Display: 29

On Private Property: 71

Square Tooth Design: 87

Round Tooth Design: 13

Size

23-24" diameter, 29-30" long
Exceptions: 2 in Steinbach 24D" x 34" long, and in 1 Medicine Hat, unmeasured. One in Newton 23D" x 28" long.
One in Topeka at the Kansas State Historical Society is 22"D x 28" long. One in North Dakota, unmeasured.

Weight

I am confused about the weight, many authors suggest a weight of about 400 pounds and some report a range from 360-720 pounds. Yet I have the weights of 3 stones, weighing in at 720, 760, and 820 each, all heavier than the prior authors suggest.

Tapered

Many writings talk of a tapered stone, but virtually all are essentially cylindrical and not tapered. Two show obvious taper, one in Buhler and one in Kansas City.

No Axle Hole

It is obvious that the stones without axle holes were never used. Speculation is that they were obsolete before the orders were complete, so they were sold as lawn ornaments and novelties. I found 13 with no axle hole.

Square Axle Hole

I found 5 with square holes. This indicates that the stone was fixed to the axle so that it would not turn separately from the stone. This was usually not a through hole.

Salt Lick

I found 10 threshing stones that have been re-purposed with one end hollowed out to be used for a salt lick stand in the corral. Some of these had axle holes and some did not. Often the sharp edges of the teeth would be smoothed off to protect the neck of the cow.

Threshing stone in German Black Sea village, Ukraine. 2008
Photo - Merv Weiss

German village of Selz, Odessa district, about 40 miles northwest of Odessa city. 2008
Photo - Merv Weiss

Karolina sitting on stone in Beresan district of Nikolaev, Ukraine. 2008
Photo - Merv Weiss

Threshing stone in the vicinity of Makarovka. May 2009
Photo - Alexander Spack

Threshing stone demonstration - State M. A. Sholokhov Museum-Reserve in Stanitsa Veshenskaya Ukraine. Public Domain

Soviet village of Lower Krynk website http://500ru.ru/pamy-at/istoriya3/index.htm.

Soviet village of Lower Krynk website http://500ru.ru/pamy-at/istoriya3/index.htm.

Soviet village of Lower Krynk website http://500ru.ru/pamy-at/istoriya3/index.htm.

Soviet village of Lower Krynk website http://500ru.ru/pamy-at/istoriya3/index.htm.

Extreme Crimea http://extreme.crimea.ua/node/847

Many of the salt lick stones show extreme deterioration due to the corrosiveness of the salt on the limestone. I assume the weight of the stone was a deterrent to being overturned, thereby keeping the salt up and out of the dirt and mud - a good solution for an item that was otherwise obsolete.

Rein Ties
3 stones have rings for tying up horse reins.

Cracked Stones
I found 6 threshing stones with cracks. Some have been glued back together and others like ours is now in 3 pieces.

Displayed On Side: 71

Displayed On End: 29
Two of these were used to hold a bird bath and two were used to hold a flag pole.

Towing Frames
All the original wooden frames have long rotted away, but I suspect that the majority were made of rough hewn wooden beams. Only the steel frames survive today. In Kansas we seem to be most familiar with the design where there is only a front cross-bar and two side rails.

Conclusion

I have probably made a big and complicated story out of a small and simple tool. The old threshing stones are obscure to most of the world, but well known to many of us, and now we even know a bit more than we used to.

References

Article 1: Threshing Stone Basics
1. D. B. Grigg, *Agricultural Systems of the World*, (Cambridge: Cambridge University Press, 1989), 53.
2. V. Gorlenko, "Types of Economic Activity," access date 9.8.12, http://etno.uaweb.org/mynuvshyna/r03html.
3. Wesley J. Prieb, Director of the Center for Mennonite Brethren Studies, Tabor College, Hillsboro, Kansas, "Memories of Rev. John K. Siemens," retired in Hillsboro, Kansas as referenced in a flier written for the Kansas Wheat Centennial - 1974.
4. Goessel Mennonite Heritage and Agricultural Museum, Note by wood shovel, "Wooden grain shovels were carved from a single piece of wood. They were used for moving grain in the 19th century when farmers believed metal would bruise the seed."
5. A. N. Mink, "Lower Dobrinka," *Historical and Geographical Dictionary of the Saratov Province* (Saratov, Russia, 1898), 231-234.
6. Heinrich Goerz, *The Molotschna Settlement Winnipeg*, (Winnipeg: CMBC Publications and Manitoba Mennonite Historical Society, 1993), 54.
7. David C. Wedel, *The Story of Alexanderwohl*, Profile of a Heritage, (Goessel, Kansas: Goessel Centennial Committee, Mennonite Press, 1974), 42.

Article 2: Cereal Grains
1. Kirby Brumfield, *This Was Wheat Farming*, (Seattle: Superior Publishing Company, 1968), 11.
2. "The Natural History of Wheat Encyclopedia of Food and Culture," Gale Encyclopedia of Food & Culture access date 9.8.12. http://www. answers.com/topic/the-natural-history-of-wheat
3. "Wheat ancient and ageless," Utah State University Cooperative Extension, access date 9.8.12, http://extension.usu.edu/aitc/lessons/pdf/wheat_ancient.pdf.
4. Brumfield, 11.
5. Ibid.
6. Ibid.
7. "Story of Farming," access date 9.8.12, http://www.historylink101.com/lessons/farm-city/story-of-farming.htm.
8. "Oklahoma Ag in the Classroom: Wheat Facts," Oklahoma State University, access date 9.8.12, http://oklahoma4h.okstate.edu/aitc/lessons/extras/facts/wheat.html.
9. Ibid.
10. "The Natural History of Wheat Encyclopedia of Food and Culture," *Gale Encyclopedia of Food & Culture*, access date 9.8.12, http://www. answers.com/topic/the-natural-history-of-wheat
11. "Story of Farming," access date 9.8.12, http://www.historylink101.com/lessons/farm-city/story-of-farming.htm.
12. Fred H. Wight, *Manner And Customs of Bible Lands*, (Moody Pres, 1953), chapter 19.
13. "Wheat ancient and ageless," Utah State University Cooperative Extension, access date 9.8.12, http://extension.usu.edu/aitc/lessons/pdf/wheat_ancient.pdf.
14. Ibid.

Article 3: The History of Threshing
1. "Thresh," (Oxford Dictionary). Access date 9.8.12, http://oxforddictionaries.com/definition/english/thresh
2. R. Douglas Hurt, *American Farm Tools, from Hand-Power to Steam-Power*, (Manhattan, Kansas: Sunflower University Press, 1982), 67.
3. "Flail - Based on Latin flagellum, whip." *Random House Dictionary*, 2012. access date 9.8.12, http://oxforddictionaries.com/definition/english/flail
4. Graeme Quick and Wesley Buchele, *The Grain Harvesters*, (St. Joseph, Michigan: American Society of Agricultural Engineers, 1987), Chapter 1.
5. Ibid.
6. Ibid.
6. Ibid.
7. Thomas D. Isern, *Bull Thresher & Bindlestiffs*, (Lawrence, Kansas: University Press of Kansas, 1990), 13-14.
8. "The Threshing Board," Access date 2010, http://vasatwiki.icrisat.org/index.php/History_of_threshing.
9. Museum of Archaeology and Evolutionary Ecology, Volgograd, access date 9.8.12, http://museum.vgi.volsu.ru/index.php/2010-03-05-06-51-05/87-2010-03-05-06-49-49&ei-Threshing Stone.
10. George Washington's Mount Vernon Estate, Museum & Gardens, access date 9.8.12, http://www.mountvernon.org/content/sixteen-sided-barn.
11. J.C. Loudon, *Encyclopedia of Agriculture*, (London: Longman, Rees, Orme, Brown, and Green, 1831), 49.
12. Bridget Jolly, "Sketch notes on South Australia's Onkaparinga threshing roller, and some antecedents," access date 9.8.12, http://www.sahistorians.org.au/175/bm.doc/threshing-roller-part-2.pdf.
13. Ibid.
14. Ibid.
15. Thomson Gale, "Threshing Machine from World of Invention," 2006, access date 9.8.12, http://www.bookrags.com/research/threshing-machine-woi.
16. Ibid.
17. Ibid.
18. "When Threshing Machines were Harvest Kings," (Oswego, Illinois: Leger-Sentinel, August 2, 2007), access date 9.8.12, http://www.ledgersentinel.com/article.asp?a=6393.
19. Christopher Wiley, "Combine Harvester: Innovating Modern Wheat Farming, Impacting the Way the World Thinks About Bread", History Link File #9483, access date 9.8.12, http://www.historylink.org/index.cfm?DisplayPage=output.cfm&file_id=
20. Wally Kroeker, "Marking Time," (Historical Committee & Archives of the Mennonite Church: Mennonite Historical Bulletin, 1999).
21. Wally Kroeker, "Farm Machinery," *Global Anabaptist Encyclopedia Online*, (1956), access date 9.8.12, http://www.gameo.org/encyclopedia/contents/F374ME.html.
22. C. Henry Smith, *Story of the Mennonites*, (Newton, Kansas: Faith and Life Press, 1981), 305.
23. "Gleaner: 85 Years of Harvest History", *Gleaner Agco Company*, access date 9.8.12, http://www.agcoiron.com/fileUpload/GLEANER_85Years_Brochure.pdf.

Article 4: The Stone and Geology
1. A. N. Mink, *Historical and Geographical Dictionary of the Saratov Province* (Saratov, Russia, 1898), 231-234.
2. William B. Bracke, *Wheat Country*, (New York: Duell, Sloan & Pearce, 1950), 90.
3. Interview with Roland Sawatzky, Senior Curator at Mennonite Heritage Village, Stein-

bach, Manitoba.

4. David Grisafe and Rex Buchanan, "Kansas Limestone", *Kansas Magazine*, (3rd Issue, 1979), 5-6.

5. "Geofacts, from the Kansas Geological Survey, Flint Hills: Rocks and Minerals," access date 9.8.12, http://www.kgs.ku.edu /Extension/flinthills/rocks.html

6. Raymond C. Moore, John C. Frye, John Mark Jewett, Wallace Lee, and Howard G. O'Connor, *The Kansas Rock Column*, (The University of Kansas, State Geological Survey, Bulletin 89 - 1951)

7. David A. Grisafe, *Kansas Building Limestone, Mineral Resources Series 4*, (Lawrence, Kansas: Kansas Geological Survey University of Kansas, 1976), 5-6.

8. "Kansas State Engineer" – Vol. 38 1956-1957 - January 1957, 27.

9. Grisafe, 16.

10. "Geology and the Prairie,"National Park Service, U.S. Department of the Interior, Tallgrass Prairie National Preserve, accesse date 9.8.12 http://www.nps.gov/tapr/upload/Geology%20brochureFINAL.pdf

11. J.G. Haskell, "Stones of Kansas," *Report to the Commissioners of Agriculture for the Year 1873*, (Washington D.C.: Washington Government Printing Office, 1874), 336.

12. Ibid.

13. Will Gilliland, Adjunct Professor at Washburn State University and retired Environmental Scientist for the State of Kansas, author's interview with Mr. Gilliland - October 2011.

14. Grisafe, 4.

15. "To the Editor of the Commonwealth," August 10, 1871, access date 9.8.12, http://www.ausbcomp.com /~bbott/cowley/oldnews/papersup/871_06com.htm

16. Sandra Van Meter, *Marion County Kansas Past and Present*, (Hillsboro, Kansas: Marion County Historical Society, Inc., M. B. Publishing House, 1972), 150.

17. Ibid.

18. Ibid.

19. Ibid.

Article 5: Making: Who-How-Where

1. A. N. Mink, *Historical and Geographical Dictionary of the Saratov Province* (Saratov, Russia, 1898), 231-234.

2. William B. Bracke, *Wheat Country*, (New York: Duell, Sloan & Pearce, 1950), 90.

3. David V. Wiebe, *They Seek a Country*, (Place: Pine Hill Press, 1974), 55.

4. "Threshers," *Newton Kansan*, April 22, 1875.

5. David A. Grisafe, *Kansas Building Limestone, Mineral Resources Series 4*, (Lawrence, Kansas: Kansas Geological Survey University of Kansas, 1976), 16.

6. Interview with Norman Epp, limestone sculptor, March 2010.

7. United States Department of the Interior - National Park Service, National Register of Historic Places Continuation Sheet Section number 7 page 1, "Bichet School District 34."

8. Note on back of Threshing Stone drawing, Chuck Schmidt owner.

9. David C. Wedel, *The History of Alexanderwohl*, (Goessel, Kansas: Goessel Centennial Committee, 1974), 42.

10. Interview with Dr. Paul Johnston, Professor Emeritus of Emporia State University, on several occasions.

11. Interview with Will Gilliland, Adjunct Professor at Washburn State University and retired Environmental Scientist for the State of Kansas, October 2011.

Article 6: Design - Why Seven Ridges

1. Wesley J. Prieb, "Trunks, Trails and Threshing Stones," Direction 3.3, October 1974, 249-52, access date 9.8.12, http://www. irectionjournal.org/article/?139

Article 7: German/Russian/Mennonite History

1. Harley Stucky, *A Century of Russian Mennonite History in America*, (Place: Swiss Mennonite Cultural & Historical Association, 1974), 2.

2. Kendal Bailes, "The Mennonites Come to Kansas," (American Heritage, magazine, August 1959, access date 9.8.12, http://www.americanheritage.com /content/mennonites-come-kansas

3. Cornelius J. Dyck, *An Introduction to Mennonite History*, (Scottdale, Pennsylvania: Herold Press, 1967), 9.

4. "Simons, Menno (1496-1561)," *Global Anabaptist Mennonite Encyclopedia Online*, access date, 25 August 2012, http://www.gameo.org/encyclopedia/contents/simons_menno_1496_1561.

5. Bailes.

6. Ibid.

7. Dyck.

8. Henry B. Tiessen, *The Molotschna Colony, A Heritage Remembered*, (Kitchener, Ontario: Henry B. Tiessen, 1979), 7.

9. Cornelius Krahn, *From the Steppes to the Prairies* – 1874 -1949, (Newton, Kansas: The Historical Committee of the General Conference of the Mennonite Church of North America, Mennonite Publication Office, 1949).

10. Hugh P. Coultis, *The Introduction and Development of Hard Red Winter Wheat in Kansas*, (Place: Report of the Kansas State Board of Agriculture 39, 1920), 217.

11. Krahn, 2.

12. Edmund G. Kaufman, *General Conference Mennonite Pioneers*, (North Newton, Kansas: Bethel College, 1973), 77.

13. Coultis, 217.

14. Dyck, 137.

15. Norman E. Saul, "The Migration of the Russian-Germans to Kansas," *Kansas Historical Quarterly*, Vol. 40.1, 1974, 38-62.

16. Krahn, 3.

17. Ibid.

18. Ibid.

19. Noble L. Prentis, *Kansas Miscellanies*, (Topeka: Kansas Publishing House, 1889), 163.

20. Krahn, 6.

Article 8: The Railroad Influence

1. "Western Argus Newspaper," February 13, 1860, 3.

2. Joseph W. Snell and Don W. Wilson "The Birth of The Atchison, Topeka and Santa Fe Railroad," *Kansas Collection: Kansas Historical Quarterly*, Vol 34.2, Summer, 1968, 113-142

3. "Private Laws of the Territory of Kansas," Passed at the Fourth Session of the Legislative Assembly; Published by Authority, (S. W. Dreggs & Co., 1859), 58.

4 Norman E. Saul, "The Migration of the Russian-Germans to Kansas," *Kansas Historical Quarterly*, Vol. 40.1, 1974, 38-62.

5. *The Topeka Commonwealth*, "End of the Track," May 20, 1869.

6. L. M. Hurley, "Newton, Kansas, a Railroad Town: History, Facilities and Operations, 1871-1971," (Mennonite Press, 1985), 1.

7. Ibid.

8. Snell, 38-62.

9. Robert Collins, *Kansas, 1874, Triumphs, Tragedies, and Transitions*, (Andover, Kansas: Robert Collins, 2011), 23-4.

10. Sandra Van Meter, *Marion County Kansas Past and Present*, (Hillsboro, Kansas: Board of Directors of the Marion County Historical Society, Inc., MB Publishing House, 1972), 39.

11. Kendal Bailes, "The Mennonites Come to Kansas," (American Heritage, magazine, August 1959, access date 9.8.12, http://www.americanheritage.com /content/mennonites-come-kansas

12. Saul.

13. Bailes.

14. Ibid.

15. Van Meter, 39.

16. *Report of the Kansas State Board of Agriculture*, 39.155-156, 947.

17. "Mennonite Life," April, 1970, 14.

18. "Topeka Commonwealth," October 15, 1874.

19. C. Henry Smith, *The story of the Mennonites*, (Newton, Kansas: Faith and Life Press, 1982), 431.

20. Van Meter, 39.

21. Bailes.

22. Bailes.

23. Saul.

24. Saul.

25. Stucky, Harley J. "Harvey County (Kansas, USA)." *Global Anabaptist Mennonite Encyclopedia Online*, 1956. Accessed 11 May 2011, http://www.gameo.org/encyclopedia/contents/harvey_county_kansas.

26. Noble L. Prentis, *Kansas Miscellanies*, (Topeka: Kansas Publishing House, 1889), 155.

Article 9: Picking Kansas

1. James Malin, "In Commemoration of the Centennial Anniversary of the Admission of Kansas into the Union 1861: An Essay to Accompany an Exhibition of the Kansas Statehood Centennial," (Place, Publisher, 1961), p#.

2 Mil Penner, *A Century on a Family Farm - Section 27*, (Place: University Press of Kansas, 2002).

3. "Kansas Archeology Basics," *Kansas Historical Society*, access date, http://www.kshs.org/p/kansas-archeology-basics/14588.

4. Ibid.

5. John D. Reynolds and William B. Lees, Ph.D. "The Archeological Heritage of Kansas," (Topeka: Kansas State Historical Society, 2004), 15.

6. "Kansas Archeology Basics."

7. Ibid.

8. Ibid.

9. "Louisiana Purchase," Library of Congress, access date, http://www.loc.gov/rr/program/bib/ourdocs/Louisiana.html.

10. "Kansas-Nebraska Act, 1854," *Our Documents*, access date, ourdocutments.gov.

11. Craig Miner, *Kansas: The History of the Sunflower State*, (Place: University Press of Kansas, 2002), 33.

12. Michael J. Merchant, *Emigrant Indian Tribes of Kansas*, (Emporia State University: Center for Great Plains Studies, 1994).

13. *A Home for Immigrants - Agricultural, Mineral and Commercial Resources of the State - Great Inducements Offered to Persons Desiring Homes in a New County - The Homestead Law 1863*, (The State of Kansas - 1865), 4-5.

14. Sandra Van Meter, *Marion County Kansas Past and Present*, (Hillsboro, Kansas: Board of Directors of the Marion County Historical Society, Inc., M. B. Publishing House, 1972), 46.

15. Ibid.

16. Wilmer A. Harms, *Our Heritage Through the Reformation, Menno Simons, the Migrations, and the Genealogy of Gerhard Theissen*, (Halstead, Kansas: Harms, 1984).

17. Ibid.

18. Bailes, "The Mennonites Come to Kansas."

20. Cornelius Krahn, *From the Steppes to the Prairies - 1874-1949*, (The Historical Committee of the General conference of the Mennonite Church of North America, Mennonite Publication Office, 1949), 7.

21. Bailes, "The Mennonites Come to Kansas."

22. Norman E. Saul, "The Migration of the Russian-Germans to Kansas," *Kansas Historical Quarterly* 40.1, 1974. access date, http://www.kshs.org/p/kansas-historical-quarterly-the-migration-of-the-russian-germans-to-kansas/13242

23. Ibid.

Article 10: Farming With Horsepower

1. Eugene A. Abalone et. al, *Marks' Standard Handbook for Mechanical Engineers 11th Edition*, (New York: McGraw Hill, 2007), 9.4.

2. Michael W. Shank, "Community Historian's Annual Number Eleven," Schaff Library, Lancaster Theological Seminary, *Community Historians* 11.6 (1972), 2.

3. Ibid.

4. Stevenson, Fletcher, *Pennsylvania Agriculture and Country Life*, 1840-1940, (Harrisburg: Pennsylvania Historical and Museum commission, 1955), 46.

5. Shank, 2.

6. Ibid, 3.

7. "Wheat: Field to Market," (Chicago: Wheat Flour Institute, 1969).

8. Mark Alfred Carleton, *Hard Wheat Winning Their Way*, (Bureau of Plant Industry, from the Yearbook of Department of Agriculture for - 1914), 400.

9. George Becker, interview with the author November 2011.

10. Kendal Bailes, "The Mennonites Come to Kansas."

11. Henry B. Tiessen, *The Molotschna Colony, A Heritage Remembered*, (Kitchener, Ontario: Henry B. Tiessen, 1979).

12. Cutler J. Cleveland, Robert K. Kaufman, *Environmental Science*, (Boston: McGraw-Hill, 2008), 318–319.

13. National Public Radio, "Here and Now, Interview with Gene Logsdon" May 11, 2012.

14. "How Much Fertilizer Do Your Animals Produce?" University of Wisconson Estension, accessed 9.8.12, http://learningstore.uwex.edu/assets/pdfs/a3601.pdf

15. James E. Beuerlein, "Wheat Growth Stages and Associated Management," (Ohio State University Extension Fact Sheet, 2001).

16. Cornelius Krahn, "Agriculture among the Mennonites of Russia," *Global Anabaptist Mennonite Encyclopedia* (1955), accessed 10 May 2011. http://www.gameo.org/encyclopedia/contents/A4128.html.

Article 11: First Person Accounts

1. Lawrence Klippenstein, Russian Revolution and Civil War. Global Anabaptist Mennonite Encyclopedia Online. 1989. Web. 10 May 2012. http://www.gameo.org/encyclopedia/contents/R871ME.html.

Article 12: Turkey Red Wheat

1. "Wheat: Field to Market," *The Kansas Wheat Commission*, (Chicago: Wheat Flour Institute, 1969),8.

2. Saul.

3. Ferenc Morton Szasz, editor, "Great Mysteries of the West," Collection of Essays: Chapter 10 by Michael L. Olsen, "And a Child Shall Lead Them, The Legendary Introduction of Turkey Red Wheat into Kansas," (Fulcrum Publishing, 1993), 169.

4. Herbert F. Friesen, *History of Turkey Hard Wheats in U.S.A.*, (Inman, Kansas: Friesen,

1974), 3.

5. J. Allen Clark, John H. Martin, and Carleton R. Ball, "Classification of American Wheat Varieties," (Washington D.C.: US Department of Agriculture Bulletin,1923), 145.

6. Karl S. Quisenberry and L. P Reitz, *Turkey Wheat: the Cornerstone of an Empire*, 1974, 98-114.

7. Ibid.

8. Saul.

9. Ibid.

10. *Marion County Record*, Marion, Kansas - November 21, 1874.

11. Karl S. Quisenberry and L. P Reitz, "Turkey Wheat: the Cornerstone of an Empire," *Agricultural History*, 48.10, 1974, 102, referencing David V. Wiebe; Cornelius Krahn; and James Malin.

12. Ibid, 103.

13. Raymond F. Wiebe, *Hillsboro Kansas, The City on the Prairie*, (Hillsboro, Kansas: Multi Business Press, 1985), 139.

14. Olsen, 165.

15. Quisenberry, 102.

16. Wiebe, 140.

17. Kansas State Network of the Mutual Broadcasting System, Dramatic reading on a radio show produced in 1941.

18. "Kansas Department of Agriculture," Kansas State Historical Society, access date, www.kshs.org/Research/Kansapedia/Theme.

19. Norman E. Saul, "Myth and History: Turkey Red Wheat and the Kansas Miracle," Heritage of the Great Plains 22.3, (1989), 6.

20. Olsen, 170.

21. Paul de Kruif, *Hunger Fighters*, (University of California: Harcourt, Brace and Company ,1928).

22. Olsen, 172.

23. James C. Milan, *Winter Wheat – in the Golden Belt of Kansas*, (Lawrence, Kansas: University of Kansas Press, 1944) 202.

24. Ibid, 250.

25. Ibid, 189.

26. Cornelius Krahn, and Richard D. Thiessen. "Warkentin, Bernhard (1847-1908)." *Global Anabaptist Mennonite Encyclopedia Online*, June 2007. Accessed 29 August 2012. http://www.gameo.org/encyclopedia/contents/W375022.html.

27. Olsen, 173.

28. Quisenberry, 104.

29. Ibid.

30. Ibid, 105.

31. Ibid. 106.

32. Walter Ebeling, *The Fruited Pane*, (Berkeley: University of California Press, 1979), 234.

33. Transcript of a dramatic reading on a radio show produced in 1941 over the Kansas State Network of the Mutual Broadcasting System.

34. L.A. Fritz, "Some Kansas Milling History," Report of the Kansas State Board of Agriculture, 1920, iv.

35. "How much wheat is produced in Kansas?" Kansas Department of Agriculture, date accessed, www.ksda.gov/kansas_agriculture/faq/id/56.

Article 13: Symbolism and Bethel College

1. "Bethel College Monthly," 8.6, June 1903, 24.

2. Peter J. Wedel, *Story of Bethel College*, (North Newton, Kansas : Bethel College, 1954), 482.

3. John Thiesen, "What's a Thresher," Bethel College Faculty Lecture 2007, referencing the "Story of Our New Emblem," *Bethel College Bulletin*, March 1935, 16.

4. "Graymaroon," (Bethel College – 1936), 4-5.

5. "Bethel College Bulletin," (Bethel College - June 1957).

6. Ken Hiebert, "What is a Thresher," access date, www.bethelks.edu/bc/aboutbc/thresher.php.

7. Ibid.

Article 15: Food-Amazing Grain of Wheat

1. "Oklahoma Ag in the Classroom: Wheat Facts," Oklahoma State University, access date 9.8.12, http://oklahoma4h.okstate.edu/aitc/lessons/extras/facts/wheat.html.

2. Ibid.

3. Ibid.

4. "Wheat Info: Fast Facts," National Association of Wheat Growers access date, www.wheatworld.org/wheat-info/fast-facts.

5. "Oklahoma Ag in the Classroom: Wheat Facts."

6. "Facts about Bread," Organic Wheat Products, access date, http://www.organicwheat-products.com/?page_id=662.

7. "Oklahoma Ag in the Classroom: Wheat Facts."

8. Michael Chu, "Wheat Flour, Kitchen Notes, Cooking For Engineers," access date, www.cookingforengineers.com/article/63/Wheat-Flour.

9. "Facts about Bread," Organic Wheat Products, access date, http://www.organicwheat-products.com/?page_id=662.

10. "Oklahoma Ag in the Classroom: Wheat Facts"

11. "When Food Changed history: The French Revolution," Smithsonian.com, access date, http:www//blogs.smithsonianmag.com.

12. "Oklahoma Ag in the Classroom: Wheat Facts"

13. Chu.

14. Ibid.

15. "Oklahoma Ag in the Classroom: Wheat Facts"

16-21. Ibid.

22. "Definition of grain," Oxford Online Dictionary, access date, www.oxforddictionaries.com/definition/english/grain.

23. Norma Jost Voth, *Mennonite Foods & Folkways - From South Russia 1.1*, (Intercourse, PA: Good Books, 1990), 36.

24. Ruth Unruh and Jan Schmidt, *From Pluma Moos to Pie*, (Goessel, Kansas: Mennonite Immigrant Historical Foundation, 1981), 68.

25. Voth, 75.

26. Ibid.

27. Ibid.

28. Ibid.

29. Margaret Dick, et. al., *Off the Mountain Lake Range*, (Mountain Lake, Minnesota: Gopher Historians, 1958), 11.

30. Ibid.

If we still used a threshing stone, would it look like this?

Author's concept model and rendering of a "Modern" Threshing Stone
Johann Deere - Power Thresher - Model 606